A Tutor to Astronomie
and Geographie

JOSEPH MOXON
1627 – 1691

Joseph Moxon

A Tutor to Astronomie
and Geographie

A FACSIMILE OF THE
FIRST EDITION
1659

RENASCENT BOOKS

PUBLICATION HISTORY

A TUTOR TO ASTRONOMIE AND GEOGRAPHIE
JOSEPH MOXON
First Edition
1659
Subsequent Editions with Appendix on the Use of the Ptolemaic Sphere
Second Edition
1670
Third Edition
1674
Fourth Edition
1686
Fifth Edition
1698
Fifth Edition
(Corrected and Enlarged by Phillip Lea)
1699
also
A TUTOR TO ASTRONOMY AND GEOGRAPHY
OR
THE USE OF THE COPERNICAN SPHERE
1665

This facsimile edition published by
TGR Renascent Books
27 Springdale Court
Mickleover, Derby DE3 9SW
United Kingdom
2011

Paperback edition first published 2013

ISBN 978-1-4826373-3-5
www.renascentbooks.co.uk

Origination, layout and typesetting by
Gordon Roberts
Transcript and proofing by
Elizabeth Roberts

Printed and bound by
CreateSpace, Charleston, South Carolina, U.S.A.

INTRODUCTION

The first of Joseph Moxon's popular scientific books was his translation of William Bleau's *Institutio Astronomica*, which he published in 1654 as *A Tutor to Astronomy and Geography*. He evidently liked the title because he used it again for his own book, based somewhat on Bleau's material but which he authored independently and published in 1659, with the title *A Tutor to Astronomie and Geographie, or Use of both the Globes, Coelestial and Terrestrial*. It is a facsimile copy of this book which you now hold in your hands. It was quite successful and ran to four editions in Moxon's lifetime, with two more editions following after his death. All these texts described the Ptolemaic version of astronomy and geography, but in 1665 Moxon published a popular work on Copernican astronomy. Resolutely adhering to his favourite title, he confusingly called this latter work *A Tutor to Astronomy and Geography, or Use of the Copernican Sphere*.

LIFE AND TIMES

Joseph Moxon was born at Wakefield, Yorkshire, on 8 August 1627. His father, James, a Puritan who appears to have resided mainly in London from about 1622, moved to Rotterdam in 1638 in order to have the freedom to print English bibles and other puritan texts. He took his family with him and while there young Joseph, together with his elder brother James, learned the printing trade. The family eventually returned to London and by 1646 the brothers had established themselves as printers in the capital. Within three years at least twelve books appeared with their imprint, among which was *A Book of Drawing, Limning, Washing or Colouring of Mapps and Prints*. Published in 1647 for the map seller Thomas Jenner, it may have been this book that set Moxon on the path to his future career. Joseph gave up his share of the printing business and from 1650 onwards his brother continued on his own.

Now with time to spare, Joseph began to study practical mathematics and the science of globe and map making. In the spring of 1652 he visited Amsterdam, where he commissioned an engraver to cut copper globe-printing plates. Back in London, he was soon advertising celestial and terrestrial globes for sale, some of a large size nearly 18 inches in diameter. Establishing himself at the Sign of Atlas, with premises initially in Cornhill, he steadily built up a business publishing scientific books as well as printing maps, charts, globes and paper mathematical instruments. In 1665 he moved from Cornhill to Ludgate Hill, where the Sign of Atlas remained until 1686 – apart from a six-year period in Russell Street enforced by the great fire of London in 1666.

During his lifetime Moxon published more than thirty popular scientific expositions and technical handbooks. He acquired an enviable reputation for printing mathematical texts, among which were Edward Wright's *Certain Errors in Navigation* (1657) and John Dansie's *Mathematical Manual* (1654). He became noted especially for the accurate printing of tabulated data, such as his tables of solar declination *Primum Mobile* (1656), and tables of logarithms which were reprinted in John Newton's *A Help to Calculation* (1657). It was also Moxon who typeset the 230 pages of trigonometrical functions and logarithms in William Oughtred's *Trigonometria* (1657). In January 1662 he was appointed hydrographer to Charles II, '...for the making of Globes, Maps and Sea-Platts', his petition for the post being supported by no less than thirteen members of the London mathematical community, three of whom were fellows of the Royal Society (Elias Ashmole, Lawrence Rook and Walter Pope).

Moxon's own election to the Royal Society in 1678 is notable for his being the first tradesman to be so elevated, a precedent that was not repeated during the seventeenth century. However, his reputation as a printer and publisher and his position as Hydrographer Royal were long known to many society fellows. It is possible that his election was linked to the languishing program for a 'history of trades' which the society had set in motion in the 1660s. Moxon independently revived this idea by publishing the first part of his *Mechanick Exercises, or the Doctrine of Handy Works* in 1678. In a following sequence of fourteen parts, he printed illustrated accounts of the trades of smith, joiner, carpenter and turner, copies of which John Evelyn presented to the president of the Royal Society. Sadly, despite his activities on these accounts, Moxon was one of twenty-three members expelled in November 1682 for non-payment of subscriptions.

From 1662 Moxon placed 'Hydrographer to the King's Most Excellent Majesty' on the title pages of his books. To this he speedily appended 'Member of the Royal Society' after his election and, somewhat audaciously, retained the distinction even after his expulsion. Both these honorifics appear on the title page of what is probably his most famous work, *Mechanick Exercises... Applied to the Art of Printing*, which appeared in twenty-four parts during 1683-4. In it Moxon details all the otherwise unrecorded craft skills and printing techniques of his day, thus preserving for posterity a precise record of the printing trade seen through the eyes of a foremost practitioner.

Joseph Moxon married three times, firstly to Susan Marsden in 1648, by whom he had a daughter Susan and a son James. His first wife died in 1659 and he married secondly Hannah Cook in 1663. He married for a third time in 1668 and after about eighteen years in this marriage, around 1686, he withdrew from trade and passed on his business to his son James. He died in February 1691 and was buried in St Paul's Churchyard.

ABOUT THIS BOOK

A Tutor to Astronomie and Geographie comprises six parts or 'books'. In the first book, Moxon teaches the rudiments of Ptolemaic astronomy and geography. In the next two books he shows how to use globes to solve many problems in astronomy, geography and navigation. The fourth book teaches how to solve astrological problems, an important subject in Moxon's time but today considered a pseudo-science of little merit. The fifth book deals with what Moxon calls gnomonical problems, that is, by again using the globes, finding the correct hour lines for many different types of sun-dials. The final book applies the globes to the solution of spherical triangles, a necessary skill for mariners practising the new art of celestial navigation. Knowledge of how to use globes in the solving of all these sorts of problems is a skill now largely forgotten and Moxon's treatise is a valuable historic resource on this account alone. Of course, the work may also be viewed as a simple handbook, produced as an aid to selling the celestial and terrestrial globes which Moxon was busy making and advertising at this time.

The treatise has two additional books, the first of which is a retelling of ancient and mythical stories about the origins and naming of certain constellations and stars, or what Moxon calls the 'poetical reasons' why such bodies are placed where they are in the heavens. The second additional book is of particular value today to historians of astronomy, since it is a masterly exposition of the origins and discoveries of astronomy up to the middle of the seventeenth century. It comprises much myth but also a great deal of fact, the whole providing a fascinating glimpse of these matters as understood by our forebears at the dawn of the scientific age.

THE SIGNS OF THE ZODIAC AND THE PLANETS, ETC.

Moxon assumes his readers are thoroughly familiar with the symbols used for the signs of the Zodiac, a reasonable assumption in his day. Therefore, he often prints the symbol in the text rather than the written name of the sign. This can make reading today rather difficult for those who do not instantly recognize the twelve symbols used for the zodiacal constellations, and accordingly they are shown in the table below alongside their written name.

♈	Aries	♎	Libra
♉	Taurus	♏	Scorpio
♊	Gemini	♐	Sagittarius
♋	Cancer	♑	Capricorn
♌	Leo	♒	Aquarius
♍	Virgo	♓	Pisces

The following symbols will also be encountered in the text without explanation, and again their meanings are written alongside in the table below.

☉	Sun	♃	Jupiter
☿	Mercury	♄	Saturn
♀	Venus		
☾	Moon	☊	Moon North Node
♂	Mars	☋	Moon South Node

Spellings and Punctuation

Printing conventions and spellings by the mid-seventeenth century were moving closer to our own modern-day usage. However, it was still the convention to use the terminal letter s at the end of words, as today, but the long form ſ everywhere else, for example poſſeſs (possess). In earlier times printers did not consider the letters u and v to be distinct, but merely variants of the same letter. Likewise for the letters i and j. Such usage does not occur often in this book but it is apparent, for example, in the diagram on page 12, where 'Mover' and 'Heaven' are spelt 'Mouer' and 'Heauen' respectively. Also note the spelling of 'Jupiter' as 'Iupiter'. The diagram in question was probably an old woodcut reused by Moxon, as evidenced by the almost obsolete spellings of Sun (Sunne), Moon (Moone), Saturn (Saturne), etc. Occasionally, other throwbacks to earlier spellings are encountered, an instance being Collumne (column) on page 68. In Book 5, the spellings 'plane' and 'plain' are used interchangeably, usually where 'plane' is intended.

Punctuation might appear somewhat eccentric to modern readers – for example commas are often used where we would expect a full stop. Words after commas and full stops might or might not be capitalised in a quite arbitrary manner. However, capital letters are sprinkled randomly throughout the text for no other apparent reason than that the compositor felt like doing it. No use is made of the apostrophe s, so occasionally there is no distinction between the singular and the plural, although the context usually makes it clear what the author intends.

Typographical Features

Justification of paragraphs in this book was not merely a cosmetic feature (as it is today). Early printers like Moxon would be laying out discrete pieces of movable metal type into a square wooden frame and if the frame was not completely filled, the types would move under the action of the press and smudge the wet ink. In other words, each line of each paragraph had to extend fully from left to right, with the letters 'jammed' in the frame or the page would be unprintable. Hence the use of aggressive hyphenation, which to a modern reader often seems surprising – for example the word 'Equinoctial' hyphenated after the first letter (E-quinoctial) with the E- on the end of a line and the rest of the word following on the next. It was also an early printing convention to follow numbers with dots, or full stops, ostensibly to distinguish counting numbers from the cardinal numbers used to represent words like first, second, third, etc. The compositors of this book have made much use of dots (and sometimes mistakenly used a comma instead of a dot). An example is found on page 45 where we find 33, degrees, 20. Minutes.

Another interesting early printing convention is found on page 30, second paragraph, where the word 'World' has the letter V used twice to represent W (VVorld). Such usage is not uncommon in printed books of the period, where it is often used far more extensively than in this present text.

Faults

The attentive reader will find some faults and errors in the pages of this book. For example, the running headings at the top of the pages in Book 5 are not all in the correct sequence. Typographical mistakes occur, as on page 73, last paragraph, where 'South Pole' is spelt 'South Nole'. The diagram on page 170 has the number III placed incorrectly on the substyle line, which throws out the following hour-line numbering. Occasional mathematical mistakes occur, as on page 22, third paragraph, where the ratio 21:8 is $2\frac{5}{8}$ and not $\frac{5}{8}$ as stated. When these and other typographical and/or mathematical errors are encountered in this book, please remember that they occur in the original printing, that no attempt has been made to indicate or remedy them and therefore they are faithfully reproduced in this facsimile reprint.

HERE BEGINS
A TUTOR TO ASTRONOMIE
AND GEOGRAPHIE

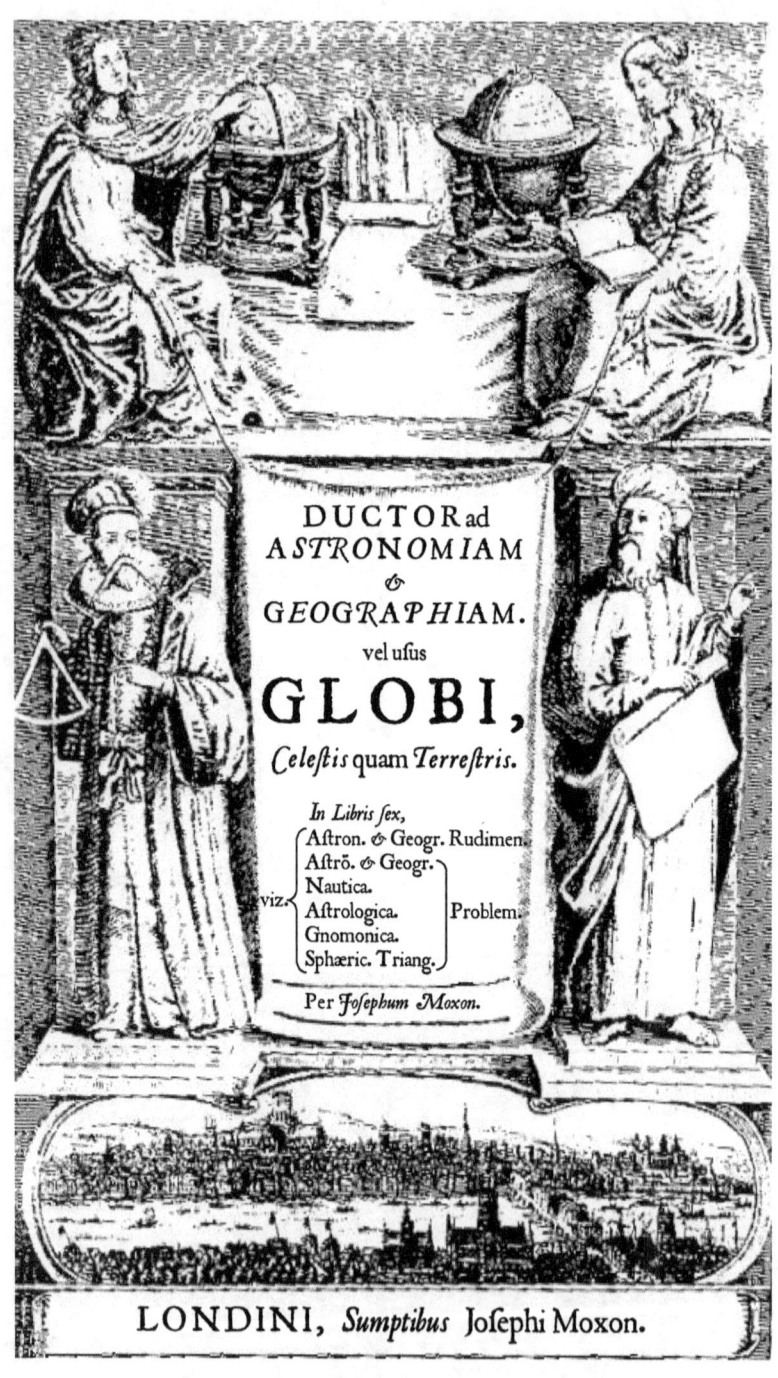

DUCTOR ad
ASTRONOMIAM
&
GEOGRAPHIAM.
vel usus
GLOBI,
Celestis quam Terrestris.

In Libris sex,
viz.
⎧ Astron. & Geogr. Rudimen.
⎪ Astrō. & Geogr. ⎫
⎪ Nautica. ⎬
⎨ Astrologica. ⎭ Problem.
⎪ Gnomonica.
⎩ Sphæric. Triang.

Per Josephum Moxon.

LONDINI, *Sumptibus* Josephi Moxon.

A *TUTOR* to

ASTRONOMIE and *GEOGRAPHIE* :

Or an Eafie and fpeedy way to know the
Ufe of both the

G L O B E S,

Coeleftial and *Terreftrial.*

In fix BOOKS.

The Firft teaching the Rudiments of *Aftronomy* and *Geography.*

The
{ 2. } Shewing by   { *Aftronomical* & *Geographical* Probl.
{ 3. } the Globes   { Problemes in *Navigation.*
{ 4. } the folution { *Aftrological* Problemes.
{ 5. } of   { *Gnomonical* Problemes.
{ 6. }   { Of *Spherical* Triangles.

More fully and amply then hath ever been fet forth either by *Gemma Frifius, Metius, Hues, Wright, Blaew,* or any others that have taught the Ufe of the Globes : And that fo plainly and methodically that the meaneft Capacity may at firft reading apprehend it ; and with a little Practife grow expert in thefe Divine Sciences.

By *Jofeph Moxon.*

Whereunto is added the *Antient Poetical Stories of the Stars* : fhewing Reafons why the feveral fhapes and forms are pictured on the *Coeleftial Globe.* Collected from Dr *Hood.*

As alfo a Difcourfe of the *Antiquity, Progrefs* and *Augmentation* of *Aftnonomie.*

Pfal. III. 2. *The Works of the Lord are great : fought out of them that have pleafure therein.*
Job. 26. 13. *By his Spirit he hath garnifhed the Heavens: His hand hath framed the crooked Serpent.*

LONDON, Printed by *Jofeph Moxon* : and fold at his Shop on *Corn-hill,* at the figne of *Atlas.* 1659.

A Catalogue of Books *and Inſtruments, Made and ſold by* Joſeph Moxon, *at his ſhop on* Corn-hil, *at the Signe of* Atlas.

GLobes of all ſizes; *Coeleſtial* and *Terreſtrial.*

Sphears, ⎧ *Ptolemean* ⎫
according ⎨ *Tychonean* ⎬ Syſteme
to the ⎩ *Copernican* ⎭

The *Catholick Planiſphere*, call'd *Blagrave's Mathematical Jewel;* made very exactly on Paſt-boards; about 17. inches Diameter. And a Book of the Uſe of it newly ſet forth by *I. Palmer* M. A.

The *Spiral Line.*

Gunters Quadrant and Nocturnal; Printed and paſted, &c.

Stirrups Univerſal Quadrat. Printed and Paſted, &c,

Sea-Plats, Printed on Paper, or Parchment, and Paſted on Boards.

Wrights Corrections of Errors, in the *Art of Navigation.* The third Edition, with Additions.

Vignola, or the Compleat Architect, uſeful for all *Carpenters, Maſons, Painters, Carvers,* or any Gentlemen or others that delight in rare Building.

A new Invention to raiſe Water higher then the Spring. With certain Engines to produce either Motion or Sound by the Water : very uſeful, profitable and delightful for ſuch as are addicted to rare curioſities : by *Iſaac de Caus.*

A Help to Calculation By *J. Newton.*

A *Mathematical Manuel,* ſhewing the uſe of *Napiers* bones, by *J. Danſie.*

A Tutor to *Aſtrology,* with an *Ephemeris* for the Year 1659. intended to be Annually continued, by *W. E.*

Alſo all manner of Mathematical Books, or Inſtruments, and Maps whatſoever, are ſold by the foreſaid *Joſeph Moxon.*

To the Reader.

Courteous Reader,

Formerly Printed a Book of the Use of the Globes, Intituled A Tutor to Aſtronomy and Geography : The Book was Compoſed by William Blaew, but the Title was mine own; and therefore I hope I may be the bolder to uſe it when and where I liſt. The ſale of that Impreſſion had almoſt perſwaded me to have Printed it again : But when I conſidered it wanted many neceſſary Problems, both in Aſtronomy, Navigation, Aſtrology, Dyalling, and the whole Doctrine of Triangles by the Globe ; And alſo that the Examples throughout that Book were made for the Citty of Amſterdam ; which by the general ſale of the Book I found rendred it leſs acceptable then it would have been if they had been made for London; And when I conſidered that to add ſo many Problems, and alter all the Examples would both Metamorphoſe that Book, and be as Laborious a work to me as if I ſhould write a new one; Then I reſolved to take this Task upon me ; which at length with Gods Aſſiſtance I have finiſhed ; And now expoſe it to thy acceptance.

The Globes is the firſt Studie a Learner ought to undertake : for without a competent knowledge therein he will never be able to underſtand any Author either in Aſtronomy, Aſtrology, Navigation, or Trigonometry : Therefore my aim hath been to make the Uſe of them very plain and eaſie to the meaneſt Capacities : In proſecution of which Deſigne, I doubt the Learneder ſort may be apt to Cenſure me guilty of Prolixity, if not Tautology : Becauſe the Precepts being plain, they may account ſome of the Examples Uſeleſs. But I deſire them to conſider that I write not to expert Practitioners, but to Learners ; to whom Examples may prove more Inſtructive then Precepts. Beſides, I hope to encourage thoſe by an ample liberal plainneſs to fall in love with theſe Studies, that formerly have been diſheartned by the Crabbed brevity of thoſe Authors that have (in Characters as it were) rather writ Notes for their own Memories, then ſufficient Documents for their Readers Inſtructions.

The

To the Reader.

The Globes for which this Book is written are the Globes I set forth about four years ago : which as I told you in my Epistle to the Reader of Blaew's *Book differs somewhat from other Globes, and that both the* Coelestial *and the* Terrestrial *; mine being the latest done of any, and to the accomplishing of which, I have not only had the help of all or most of the best of other Globes, Maps, Plats; and Sea-drafts, of New Discoveries that were then extant, for the* Terrestrial Globe, *but also the Advice and directions of divers learned and able Mathematicians both in* England *and* Holland *for Tables and Calculations both of Lines and Stars for the* Coelestial *: upon which Globe I have placed every Star that was observed by* Tycho Brahe *one degree of Longitude farther in the* Ecliptick *then they are on any other Globes : So that whereas on other Globes the places of the Stars were correspondent with their places in Heaven 58. Years ago, when* Tycho *observed them, and therefore according to his Rule want about 47. minutes of their true places in Heaven at this Time : I have set every Star one degree farther in the* Ecliptick, *and Rectified them on the Globe according to the true place they will have in Heaven in the Year 1671.*

On the Terrestrial Globe *I have inserted all the New Discoveries that have been made, either by our own or Forraigne Navigators, and that both in the East, West, North, and South, parts of the Earth. In the* East Indies *we have by these later Times many spacious Places discovered, many Ilands inserted, and generally the whole Draft of the Country rectified and amended, even to the Coast of* China, Japan, Giloli, *&c. In the* South Sea *between the* East *and* West Indies *are scattered many Ilands, which for the uncertain knowledge former Times had of them are either wholely left out of other Globes, or else laid down so erroneously that little of credit can be attributed unto them :* California *is found to be an Iland, though formerly supposed to be part of the main Continent, whose North West shoar was imagined to thrust it self forth close to the Coasts of* Cathaio, *and so make the supposed Straits of* Anian. *The Western Shoars of the* West Indies *are more accurately discribed then formerly, as you may see if you compare my Terrestrial Globe with the Journals of the latest Navigators : And if you compare them with other Globes you will find 5, 6, yea 7, degrees difference in Longitude, in most Places of these Coasts.* Magellanica *which heretofore was thought*

to

to be part of the South Continent called Terra Incognita *is now also found to be an Iland. All that Track of Land called* Terra Incognita *I have purposely omitted, because as yet we have no certainty whether it be Sea or Land, unless it be of some parts lately found out by the Dutch; who having a convenient Port at* Bantam *in* Java, *have from thence sent forth Ships Southwards, where they have found several very large Countries ; one whereof they have called* Hollandia Nova, *another* Zelandia Nova, *another* Anthoni van Diemans *Land; and divers others; some whereof lies near our* Antipodes; *as you may see by my Terrestrial Globe. Again, Far to the Northwards there are some New Discoveries, even within* 6. *degrees of the Pole : The Drafts to the* North Eastwards *I have laid down even as they were discribed by the Searchers of these Parts, for a Passage into the* East Indies. *And also the Discoveries of* Baffin, *Capt.* James, *and* Capt. Fox, *(our own Country men that attempted the finding a passage that way into the South Sea.*

I also told you what difference there is in several Authors about placing their first Meridian, which is the beginning of Longitude; that Ptolomy *placed it at the* Fortunate Ilands, *which* Mr Hues *pag.* 4. *chap* 1. *in his Treatise of Globes proves to be the Ilands of* Cabo Verde, *and not those now called the* Canary Ilands ; *because in his Time they were the furthest Places of the Discovered World towards the Setting of the Sun : Others placed it at* Pico *in* Teneriffa ; *Others at* Corvus *and* Flora ; *because under that Meridian the Compass had no Variation, but did then duely respect the North and South ; Others for the same Reason began their Longitude at* St Michaels ; *and others between the Ilands of* Flores *and* Fayal : *And the Spaniards of late by reason of their great Negotiation in the* West Indies, *have begun their Longitude at* Toledo *there, and contrary to all others account it Westwards.*

Therefore I seeing such diversity among all Nations, and as yet a Uniformity at home, chose with our own Country men to place my First Meridian at the Ile Gratiosa, *one of the Iles of the* Azores.

By the different placing of this first Meridian it comes to pass that the Longitude of places are diversly set down in different Tables : For those Globes or Maps that have their first Meridian placed to the Eastwards of Gratiosa *have all places counted Eastward between the first Mertdian and the Meridian of* Gratiosa

tiosa *in fewer degrees of Longitude : And those* Globes *and* Maps *that have their first* Meridian *placed to the* Westwards, *have all Places counted* Eastwards *from the* Meridian *of* Gratiosa *and their first* Meridian *in a greater number of degrees of Longitude, and that according as the* Arch *of* Difference *is.*

I have annexed a smal Collection *out of* Dr Hood, *which declares the* Reason *why such strange Figures and Forms are pictured on the* Celestial Globe *: and withall the* Poetical Stories *of every* Constellation.

I also thought good to add at the latter end of this Book *a smal* Treatise, *intituled* The Antiquity, Progress, and Augmentation of Astronomy. *I may without* Partiallity *give it the* Encomium *of a Pithy, Pleasant, and* Methodical *peece : It was written by a Learned* Author *; and is worthy the* Perusal *of all Ingenuous Lovers of these Studies.*

Joseph Moxon.

Encomiastic Achrosticon Authoris.

J Ts now since *Atlas* raign'd thousands of Years,

O F whom 'tis Fabl'd, Heavens hee did Uphold,

S O Ancient Authors write: But it appeares

E Xcell he others did, for we are told

P Roject he did the Sphear : and for his Skil

H E had therein, his Fame will Flourish still.

M Ust we not also Praise in this our Age

O Ur Authors skill, and Pains, who doth ingage

X Thousand Thanks, not for this Book alone

O Fhis, But for the *Globes* he makes there's none

N Ow extant made so perfect: This is known

The Contents

of the Firſt Book.

A   *Of*

The Contents.

The Contents

Of the Second Book.

The Contents.

A 2

The Contents.

The Contents.

The Contents
Of the Third Book.

A 3

The Contents.

The Contents
Of the Fourth Book.

The Contents.

The Contents
Of the Fifth Book.

The

The Contents.

The Contents
Of the Sixth Book.

The First BOOK.

Being the first RUDIMENTS *of*

Astronomy & Geography.

Or

A Description of the Lines, Circles, *and other* Parts *of the* G L O B E.

P R Æ F A C E.

THe Students of all *Arts and Sciences have ever proposed a Maxime, whereon (as on an allowed Truth) the whole Science hath dependance: and by so much the more demonstrable that Maxime is, so much the more of Excellency the Science may claim.*

This of Astronomy *and* Geography *comes not behind any; for herein we shall only admit (with the Ancients) that the Form of the visible World is* Spherical: *Neither shall we beg our Ascertion any farther then* Occular *Appearance will demonstrate: every Mans Ey being his Judge, if he be either on a Plain field, or at Sea, where nothing can hinder a free inspection of the* Horizon.

Vpon good grounds therefore they ascerted the Spherical form of the Whole: and also concluded the Parts to be Round: I meane, very intire Subsistence, as the Stars, Planets, *and the* Earth. *In the* Celestial Bodies *(as the* Stars *and* Planets) *this is also visible; and therefore un-controullable: But that the* Earth *is Round proves with the unskilfull matter of di-*

B
spute.

ſpute; *they frequently objecting with* S. Auſtine *the words of the Scripture, which ſay,* He hath ſtretched forth the corners of the Earth *; not conſidering whether thoſe words were ſpoken as alluding to the amplitude of Gods Omnipotence; or that the Corners were meant* Capes *of* Land, *which indeed are ſtretched forth into the Sea. But that the Earth is* Round *is proved by divers certain and infallible* Reaſons,

As firſt, By the Navigations of our Age, Divers able and honeſt Mariners having Sailed and continued an Eaſterly Courſe, have at length arrived (without turning back) to the ſame place from whence they ſet forth : witneſs Magellanicus, *Sr.* Francis Drake, Tho. Cavendiſh, Oliver vander Noort, W. Schouten, *&c.*

Secondly, By the length of degrees in every Parallel; *for it is found by Dayly obſervation that the degrees of every* Parallel *upon the Earth, hold the ſame proportion to the degrees of the Equinoctial, as the degrees of the ſame* Parallel *upon an Artificial Globe or Sphear do to the degrees of the greateſt Circle of the ſame : This Argument alone is ſufficient : yet take one more from Viſible Appearance : And that is this : The ſhadow which the Earth and Water together make in the Eclipſe of the Moon is alwaies a part of a Circle ; therefore the Earth and Water which is the Body ſhadowing muſt alſo be a Circular or round Body; for if it were three ſquare, four ſquare, or any other form, then would the ſhadow which it makes in the Moon be of the ſame faſhion.*

Beſides, Of all figures the Sphear or Globe is moſt perfect, moſt Capacious, and moſt intire of it ſelf, without either joynts or Angles; which form we may alſo perceive the Sun, Moon, *and* Stars *to have, and all other things that are bounded by themſelves, as Drops of Water, and other liquid things.*

But there is another frequent Argument againſt the Globulus *form of the Earth ; and that is, That it ſeems im-*

poſſible

poſſible that the Earth ſhould be round, and yet alſo Inhabible in all Places : For though we that inhabite on the top of the Earth go with our heads upwards; yet thoſe that inhabite underneath us muſt needs go with their Heads downwards, like Flyes on a Wall or Ceeling; and ſo be in danger of falling into the Air.

For Anſwer hereunto, firſt, You muſt underſtand that in the Center of the Earth there is an Attractive and drawing power, which draws all heavy ſubſtances to it : by vertue of which Attractive power, things though looſed from the Earth will again incline and cling to the Earth, and ſo much the more forcibly, by how much the heavier they are ; as a bullet of Lead let fall out of the Air, inclines towards the Earth far more violently and ſwiftly then a bullet of the ſame bigneſs of Wood, or Cork.

Secondly, you muſt underſtand that in reſpect of the whole Vniverſe there is no part either upper or under, but all parts of the Earth are alike incompaſt with Heaven; yet in reſpect of the Earth, it is Heaven, which we take for the upper part; and therefore we are ſaid to go with our heads upwards, becauſe our head (of all the parts of our body) is neareſt to Heaven.

Now that this Attractive power lies in the Center of the Earth, is proved by this Argument: If the Attractive power were not in the Center, a Plumb-line let fall would not make Right Angles with the Superficies of the Earth; but would eb Attracted that way the Attractive vertue lies, and ſo make unequal Angles with the Superficies : But by ſo many Experiments as hath yet been made, we find that a Plumb-line continued, though never ſo deep, yet it alters no Angles with the Superficies of the Earth; and therefore undoubtedly the Attractive power lies in the very Center, and no where elſe.

CHAP.

CHAP. I.

I. *What a* Globe *is*.

A Globe according to the Mathematical Definition, is a perfect and exact round Body contained under one ſurface.

Of this form (as hath been proved) conſiſts the Heavens and the Earth : and therefore the Ancients with much pains Study and Induſtry, endeavouring to imitate as well the imaginary as the real appearances of them both, have Invented two Globes; the one to repreſent the Heavens, with all the Conſtellations, fixed Stars, Circles, and Lines proper thereunto, which Globe is called the *Celeſtial Globe* ; and the other with all the Sea Coaſts, Havens, Rivers, Lakes, Cities, Towns, Hills, Capes, Seas, Sands, &c. as alſo the Rhumbs, Meridians, Parallels, and other Lines that ſerve to facilitate the Demoſtration of all manner of Queſtions to be performed upon the ſame : and this Globe is called the *Terreſtrial Globe*.

II. *Of the two* Poles.

Every Globe hath two Poles, the one North, the other South. The North Pole is in the North point of the Globe : The South Pole in the South point.

III. *Of the* Axis.

From the Center of the Globe both waies, proceeds a line through both the Poles, and continues it ſelf infinitely; which is called the *Axis of the World;* and is repreſented by the two wyers in the Poles of the Globe : Upon theſe two wyers the Globe is turned round, even as the Heavens is imagined to move upon the *Axis of the World*.

IIII. *Of the* Braſen Meridian.

Every Globe is hung by the *Axis* at both the Poles in a Braſen Meridian, which is divided into 360. degrees; (or which is all one) into 4 Nineties: the firſt beginning at the North Pole, is continued from the left hand towards the right till the termination of 90 degrees, and is marked with 10, 20, 30, &c. to 90. from whence the degrees are numbred with 80, 70, 60, &c. to 0. which is in the South Pole: from whence again the degrees are numbred

with

with 80, 70, 60, &c. to 0. and laftly, from 0 the degrees are numbred with 10, 20, 30, to 90. which is again in the North Pole.

This Brafen Meridian is of great ufe; for by help of it you may
find the Latitude of all Places, the Declination of all the Stars,
&c, and rectifie the Globe to any Latitude.

V. *Of the* Horizon.

The Horizon is a broad wooden Circle, encompaffing the
Globe; having two notches in it; the one in the North the other in
the South point: The notches are made juft fit to contain the Brafen Meridian that the Globe is hung in: in the bottom or under
Plane of the Horizon there ftands up a rop or (as it is called) a
Bed, in which there is alfo a notch, into which notch the Brafen
Meridian is alfo let, fo lo, as that both it and the Globe may be
divided into two equal halfs by the upper Plane of the wooden Horizon. Thefe Notches are as gages to keep the Globe from inclining more to the one fide of the wooden Horizon then the other.

Upon the upper Plane of the Horizon is feveral Circles delineated: as firft, the inner Circle, which is a Circle divided into twelve
equal parts, *viz.* into twelve *Signes;* every Signe having its name
prefixed to it; as to the Signe of ♈ is the word *Aries;* to ♉ the
word *Taurus,* &c. every Signe is again divided into 30 equal parts,
which are called Degrees, and every tenth degree is marked with
10, 20, 30.

Next to the Circle of Signes is a Kalender or Almanack, according to the Old ftile ufed by us here in *England*, each Moneth
being noted with its proper Name; as *January, February, March,*
&c. and every day diftinguifhed with Arithmetical figures, as
1, 2, 3, 4, &c. to the end of the Moneth.

The other Calender is a Calender of the New ftile; which is in a
manner all one with the Old; only in this Calender the moneth
begins ten daies fooner then they do in the other: and to this Calender (becaufe it was inftituted by the Church of *Rome*) there is
annexed the Feftival daies Celebrated by the Romifh Church.

The two other Circles are the Circles of the Winds; the inner-
moft having their Greek and Latine names; which by them were
but twelve; and the outermoft having the Englifh Nanes, which
for more precifenefs are two and thirty.

The ufe of the upper Plane of the Horizon is to diftinguifh the
Day from the Night; the Rifing and Setting of the Sun, Moon, or

Stars, &c. and for the finding the Azimuth, and Amplitude, &c.

VI. *Of the* Quadrant *of* Altitude.

The Quadrant of Altitude is a thin braſs plate, divided into 90. degrees; and marked upwards with 10, 20, 30, 40, &c. to 90. It is rivetted to a Braſs Nut, which is fitted to the Meridian; and hath a Screw in it, to ſcrew upon any degree of the Meridian. When it is uſed it is ſcrewed to the *Zenith.* Its uſe

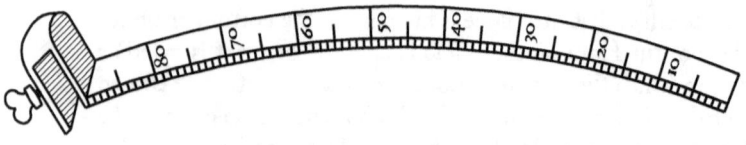

is for meaſuring the Altitudes, finding Amplitudes, and A-zimuths, and diſcribing Almicantaraths. It would ſometimes ſtand you in good ſteed if the Plate were longer by the bredth of the Horizon then 90. degrees; for then that length being turned back will ſerve you inſtead of an Index, when the Nut is ſcrewed to the Zenith, to cut either the degrees or Daies of either Style, or the Points of the Compaſs in any of thoſe Circles concentrical to the innermoſt edge of the Horizon, which the Ey cannot ſo well judge at.

VII. *Of the* Hour Circle, *and its* Index.

The Hour Circle is a ſmal Braſen Circle, fitted on the Meridi-an, whoſe Center is the Pole of the world: It is divided into the 24 hours of the Day and Night, and each hour is again divided into halfs and quarters, which in a Revolution of the Globe are all pointed at with an Index, which to that purpoſe is fitted on the Axis of the Globe.

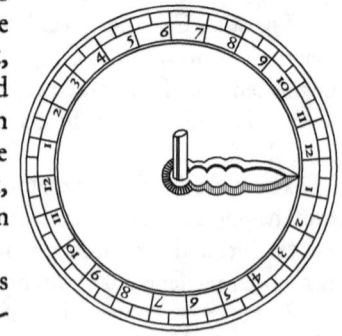

The uſe of the hour Circle is for ſhewing the Time of the ſe-veral mutations and Configurati-ons of Celeſtial Appearances.

VIII. *Of*

VIII. *Of the* Nautical Compaſs, *or* Box *and* Needle.

Juſt under the Eaſt point of the Horizon, upon the undermoſt Plane, is ſometimes fixed a *Nautical Compaſs,* whoſe North and South line muſt be Parallel to the North and South line of the Horizon. The uſe of it is for ſetting the Angles of the Globe correſpondent to the Angles of the World.

IX. *Of the* Semi-Circle *of* Poſition.

This is a Semi-Circle made of Braſs, and divided into 180. degrees, numbred from the Equinoctial on either ſide with 10, 20, 30, &c. to 90. at the two ends there is an Axis, which is fitted into the two hole, of two ſmal ſtuds fixed in the North and South points of the upper Plane of the Horizon : upon this Axis it is moved up and down, according to the intent of your operation.

The uſe of this Circle of Poſition is, for the finding the twelve Aſtrological Houſes of Heaven; and alſo for finding the Circle of Poſition of any Star or Point in Heaven.

Thus much may ſerve for the lineaments Circumjacent to the body of the Globe. The next diſcourſe ſhall be

CHAP. II.

Of the Circles, Lines, &c. *diſcribed upon the* Superficies *of the* Globe; *beginning with the* Terreſtrial Globe; *and*

I. *Of the* Equator.

THe *Equator* is a *great Circle,* encompaſſing the very middle of the Globe between the two Poles thereof, and divides it into two equal parts, the one the North part, and the other the South part. It is (as all great Circles are) divided into 360. equal parts, which are called Degrees. Upon this Circle the Longitude is numbred, from Eaſt to Weſt : and from this Circle both waies, *viz.* North and South the Latitude is reckoned. It is called the E-quator, becauſe when the Sun comes to this line (which is twice in one year, to wit, on the tenth of *March,* and the eleventh

of

of *June*) the Daies and Nights are equated, and both of one length.

II. *Of the* Meridians.

There are infinite of Meridians, for all places lying Eaft or Weft from one another have feveral Meridians; but the Meridians delineated upon the Terreftrial Globe are in number 36. fo that between two Meridians is contained ten degrees of the Equator. From the firft of thefe Meridians (which is divided into twice 90 degrees) accounted from the Equator towards either Pole) is the beginning of Longitude, which upon our Englifh Globes is at the Ile *Gratiofa*, one of the Iles of the *Azores*, and numbred in the Equator Eaftwards, with 10, 20, 30, &c. to 360. round about the Globe, till it end where it began.

They are called Meridians, becaufe they divide the Day into two equal parts: for when the Sun comes to the Meridian of any Place, it is then Midday, or full Noon.

III. *Of the* Parallels.

As the Meridians are infinite, fo are the Parallels; and as the Meridian lines delineated upon the Globe are drawn through no more then every tenth degree of the Equator, fo are the Parallels alfo delineated but upon every tenth degree of the Meridian; left the Globe fhould be too much filled with fuperfluity of lines, which might obfcure the fmal names of Places. The Parallel Circles run Eaft and Weft round about the Globe, even as the Equator; only the Equator is a great Circle; and thefe are every one lefs then other, diminifhing gradually till they end in the Pole. The Parallels are numbred upon the Meridian with 10, 20, 30, &c. to 90. beginning in the Equator, and ending in the Pole.

They are called Parallels; becaufe they are Parallel to the Equator.

IIII. *Of the* Ecliptick, Tropicks, *and* Polar Circles.

Thefe Circles though they are delineated upon the Terreftrial Globe, yet they are moft proper to the Celeftial; and therefore when I come to the Celeftial Globe, I fhall define them unto you.

V. *Of the* Rhumbs.

The Rhumbs are neither Circles nor ſtraight lines, but Heli-ſpherical or Spiral lines: They proceed from the point where we ſtand, and wind about the Globe till they come to the Pole; where at laſt they looſe themſelves. They repreſent the 32 winds of the Compaſs.

Their uſe is to ſhew the bearing of any two places one from another: that is to ſay, upon what point of the Compaſs any ſhoar or Land lies from another.

There are many of them deſcribed upon the Globe, for the better directing the ey from one ſhoar to the other, when you ſeek after the bearing of any two Lands. Some of them (where there is room for it) have the figure of the *Nautical Card* drawn about the Center or common interſection, and have (as all other Cards have) for the diſtinction of the North point, a *Flowerde-luce* pictured thereon.

They were firſt called Rumbs by the *Portugals;* and ſince uſed by Latine Authors, and therefore that name is continued by all Writers that have occaſion to ſpeak of them.

VI. *Of the* Lands, Seas, Ilands, &c. *Deſcribed upon the* Terreſtrial Globe:

The Land deſcribed upon the Globe is bounded with an irregu-lar line, which runs turning and winding into Creeks and Angles, even as the ſhoar which it repreſents (doth) For the better di-ſtinction of Lands, &c, this line is cullered cloſe by one ſide thereof with divers Cullers, as with red, yellow, green, &c. theſe cullers diſtinguiſh one part of the Continent from the other; and alſo one Iland from another. That ſide of the line which incompaſſes the Cullers, is the bounds of the Land; the other ſide of the line which is left bare without Cullers, is the limits of the Water.

The *Land* is either *Continents,* or *Ilands.*

A *Continent* is a great quantity of *Land,* not interlaced or ſepa-rated by the *Sea,* in which many *Kingdomes* and *Principalities* are contained; as *Europe, Aſia, Affrica, America.*

An *Iſland* is a part of the *Earth,* environed round with *Wa-ters* ; as *Britain, Java, S. Laurence Iſle, Barmudas,* &c.

Theſe again are ſub-divided into *Peninſula, Iſtmus, Promon-torium.* C A

A *Peninfula* is almoſt an *Iſland;* that is, a track of *Land* which being almoſt encompaſſed round with *Water,* is joyned to the firm *Land,* by ſome little *Iſtmus;* as *Molacca* in the *Eaſt-Indies,* &c.

An *Iſtmus* is a little narrow neck of *Land,* which joyneth any *Peninſula* to the Continent; as the Straits of *Dariene* in *Peru,* and *Corinth* in *Greece.*

Promontorium, is ſome high Mountain, which ſhooteth it ſelf into the *Sea,* the utmoſt end of which is called a *Cape,* as that great *Cape of Good Hope,* and *Cape Verde* in *Africa.*

The *Water* is either *Ocean, Sea, Straits, Creeks,* or *Rivers.*

The *Ocean* is that generall collection of all *Waters,* which invironeth the whole *Earth* on every ſide.

The *Sea* is a part of the *Ocean;* to which we cannot come, but through ſome *Strait,* as *Mare Mediterraneum, Mare Balticum,* and the like.

Theſe two take their names either from the adjacent places, as the Brittiſh Ocean, the *Atlantick Sea,* &c. or from the firſt diſcovere as *Mare Magellanicum; Davis,* and *Forbiſhers Staits;* &c. Or from ſome remarkable accident, as *Mare Rubrum,* from the red colour of the Sands; *Mare Ægeum, Pontus Euxinus,* and the like.

A *Strait,* is a part of the Ocean reſtrained within narrow bounds, and opening a way to the *Sea* ; as the *Straits of Gibralter, Helleſpont,* &c.

A *Creek* is a crooked ſhoar, thruſting out as it were two armes to imbrace the *Sea,* as *Sinus Adriaticus, Sinus Perſicus* &c.

A *River* is a ſmall branch of the *Sea,* flowing into the *Land;* as *Thames, Tiber, Rhine, Nilus* &c.

Now that theſe Lands, Ilands, Towns, Seas, Rivers, &c. may at the firſt ſearch be found upon the Globe, all Geographers have placed them thereon according to Longitude, and Latitude.

VII. *Longitude.*

The Longitude is an Arch of the Equator, comprehended between the firſt Meridian and the Meridian of the Place you inquire after. It is numbred on the Equator from the Weſt to the

Eaſt-

Eaftwards, with 10, 20, 30, to 360. degrees, till it end where it began.

VIII. *Latitude.*

The Latitude is an Arch of the Meridian, comprehended between the Equator and the place enquired after. It is numbred on the Meridian, from the Equator both waies, *viz.* North and South, till it come to the Poles, or 90 degrees.

Thus much may ferve for the defcription of the *Tereftrial Globe:* I therefore come to treat of the Celeftial.

CHAP. III.

Of the Celeftial Globe, *or the Eighth* Sphear, *reprefented by the* Celeftial Globe: *its motion, and of the Circles, Lines, Images, Stars, &c. defcribed thereon.*

I. *Of the eighth* Sphear.

THe eighth Sphear which is the ftarry Heaven, is reprefented by the Celeftial Globe, becaufe upon the Convexity of it all the Stars and vifible appearances are placed according to the order that they are fituated in the concavity of the eighth Sphear. It is called the *eighth Sphear,* becaufe between it and us are contained feven other Heavens, or Sphears; as 1. the *Moon,* 2. *Mercury,* 3. *Venus,* 4. *the Sun,* 5. *Mars,* 6. *Jupiter,* 7. *Saturn.* and eighthly the ftarry Heaven. The antients have made the *Syfteme* of the world to confift of 2 other *Sphears,* called the *Chriftiline Heaven,* and the *Primum Mobile,* or firft Mover: as in the following figure is reprefented.

A figure wherein may be ſeen the Compoſition of the whole frame of the World.

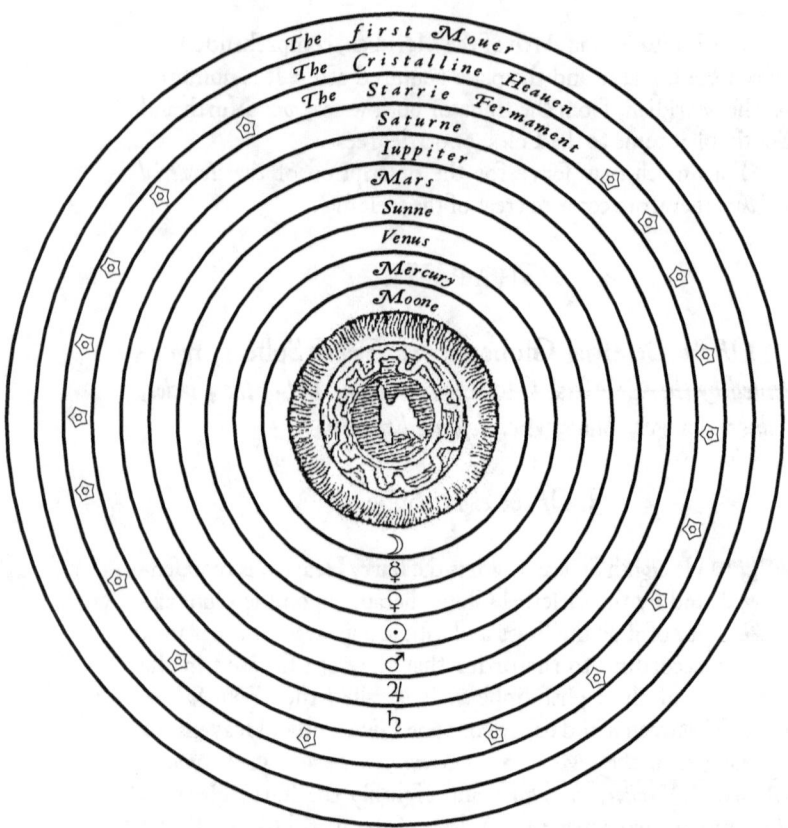

II. *Of the* *Motion of the eighth* Sphear.

There hath bin attributed to the eighth Sphear a twofold motion; the one called its *Diurnal* *Motion,* which is made from Eaſt to Weſt upon the Poles and Axis of the World: And the other called its *Second motion;* which is made from Weſt to Eaſt upon the Poles and Axis of the Ecliptick.

The *Diurnal motion* is cauſed by the violent Motion of the
Primum

Primum Mobile; for in 24 hours it carries along with it, not only the eighth Heaven or Orb of fixed Stars, but the Orbs of the *Sun,* the *Moon,* and all the reft of the Planets. It is called the *Diurnal Motion* becaufe it is finifhed in one Day.

The fecond Motion is unproperly attributed to the eighth Sphear; it being indeed the Motion of the Equinoctiall; tho Authors fometimes carelefly mention the one infteed of the other. Therefore in the next Section, where I treat of the Equinoctial, I fhall at large explain unto you the nature of this mif-called *Second Motion.*

III. *Of the* Equinoctial.

The Equinoctial upon the Celeftial Globe, is the fame line formerly called the Equator upon the Terreftrial; only with this difference, that the Equator remains fixt upon the Terreftrial Globe, but the Equinoctial upon the Celeftial Globe is moveable; (or at leaft muft be imagined to move) contrary to the *Diurnal motion* from Weft to Eaft, upon the Poles of the Ecliptick: I fay imagined to move, becaufe in the Heavens it doth really move, tho on a material Globe it would be inconvenient to make a moveable Equinoctial, and therefore it hath one fixed: which for this and the next age will fufficiently ferve, without much deviation from the truth it felf.

Now that the difference between the Equator upon the Terreftrial Globe, and the Equinoctial upon the Celeftial, may be proved; and the motion of the Equinoctial be the better underftood; I fhall only bring this example,

All places that were formerly under the Equator, do and will keep the fame Longitude, and remain ftill under the Equator: as may be proved by comparing the Ancient and modern *Geographers* together: but thofe Stars that were formerly under the E-quinoctial, do not keep the fame Longitude, nor remain under the Equinoctial: becaufe the Equinoctial (as aforefaid) hath a motion from Weft to Eaft, upon the Poles of the Ecliptick. But the Stars being fixed in their one Sphear, like knots in wood, and therefore move not, are by the Preceffion of the Equinox left behind the Equinoctial Colure, and fo are caufed to alter their Longitude; as by comparing the Obfervations of ancient and modern *Geographers* together, it will appear: for about 346

years

years before Chriſt, the firſt Star in the *Rams horn* was by the *Egyptian* and *Grecian* Aſtronomers obſerved to be in the Equinoctial Colure: and 57 years ago, when *Tycho* obſerved, it was found to be in 27 degrees 37 minutes of ♈. So that in about 2000 years it is moved forwards 28 degrees, and will according to *Tycho's* opinion, finiſh its Revolution in 25412 years : According to which motion, I have Calculated this following Table, for finding the Degrees and Minutes of the Equinoctial motion, anſwerable to any number of years within the ſaid Revolution.

ye.	deg.	m		years.	deg.	m.
1	0	$0\frac{3}{4}$		100	1	25
2	0	$1\frac{1}{2}$		200	2	50
3	0	$2\frac{1}{2}$		300	4	15
4	0	$3\frac{1}{3}$		400	5	40
5	0	$4\frac{1}{4}$		500	7	5
6	0	5		1000	14	10
7	0	$5\frac{3}{4}$		2000	28	20
8	0	$6\frac{1}{2}$		3000	42	30
9	0	$7\frac{1}{4}$		4000	56	40
10	0	$8\frac{1}{2}$		5000	70	50
20	0	17		10000	141	40
40	0	34		20000	283	20
60	0	51		25000	354	10
80	1	8		25412	360	

This Table may be of uſe for finding the Equinoctial poſition of any Star, for any year either paſt, preſent, or to come. Its uſe is very eaſie, for if you deſire to know the motion of the Equinox for any number of years, you need but ſeek your number in the Collumn of years, and againſt it you have the degrees and minutes of the Equinoctial motion.

But tho the Stars have this motion one way, *viz.* in Longitude, yet do they not at all alter their Latitudes; becauſe the motion of the Equinoctial is made upon the Poles of the *Ecliptique*.

IIII.

IIII *Of the* Ecliptique.

The *Ecliptique* is a great Circle, lying oblique or aslope from
the Equinoctial, making an Angle of 23 ½ degrees with it: It
cuts the Equinoctial into two equal parts, and is cut by the E-
quinoctial in two opposite points, *viz.* ♈, and ♎. It divides the
Globe into two equal parts, called *Hemisphears;* the one the
Northern and the other the Southern *Hemisphear.* It is divided
into 12 equal parts, which are called the twelve *Signes,* every
part being noted with the Character of the *Signes* belonging
unto it; as unto *Aries,* ♈: to *Taurus,* ♉: to *Gemini,* ♊; and so of
the rest. From every one of these 12 divisions proceed both
waies *viz.* North, and South, Circles of Longitude, into the
Poles of the *Ecliptique.* Each of these twelve *Signes* is divided
into 30 equal parts, which are called degrees; and are numbred
upon every tenth degree with 10, 20, to 30, and upon may new
Celestial Globe, for more precisenes, every degree is again divided into halfs.
„ It is called the *Ecliptique* as being derived from the Greek
„ word: Εχλεπσίν which signifies to want light, because in and
„ about it happen all the defects and Eclipses both of the Sun
„ and the Moon.
„ It is also called *the Way of the Sun,* because the Sun goes
„ alwaies under it, passing through it in all his Annual Course.

V. *Of the* Poles *of the* Ecliptick.

There are two Poles of the Ecliptick, the one the North Pole,
the other the South Pole; and are called North or South according to their position next the North or South Pole of the
World. Each it distant from its correspondent Pole of the
World 23 degrees 30 minutes.
As on the Terrestrial Globe all the Meridians discribed thereon meet in the Pole of the World, so on the Celestial all the Circles of Longitude drawn through the twelve Signes meet in the
Poles of the Ecliptick.

VI.

VI. *Of the* Axis *of the* Ecliptick.

Through the two Poles of the *Ecliptick* is imagined to paſs a ſtraight line, through the Center of the Plain of the *Ecliptick;* which is called the *Axis of the Ecliptick,* upon which the *ſecond motion* of the *Ecliptick* is performed: even as the Diurnal motion is performed upon the Axis of the World.

VII. *Of the* Colures, *and* Cardinal Points.

There are two great Circles cutting one another at right an-gles in the Poles of the World, which are called the *Colures.* Each *Colure* receives an additional name from the point in the *Ecliptick* that it Cuts; as the one paſſes from Pole to Pole through the beginning of ♈ and ♎, which being two Equinoctial Signes, name therefore that Colure the *Equinoctial Colure:* The other paſſes through the beginning of ♋ and ♑, which are Solſticial Signes, and therefore names that the *Solſticial Colure.*

These Colures by interſecting one another, divide themſelves into four Semi-circles; and theſe Semi-circles divide the *Ecliptick* into four equal parts. *viz,* in ♈, ♋, ♎, and ♑,

The points of the Ecliptick that theſe interſections paſs through, are called the *four Cardinal points,* and are of great uſe in A-ſtronomy; for according to the Suns approach to any of them, the Seaſon of the year is altered into *Spring, Summer, Autumn, Winter:* as ſhall be ſhewed hereafter.

VIII. *Of the* Tropicks.

There are two ſmaller Circles Parallel to the Equinoctial, which are called the *Tropicks;* the one called the *Tropick of Can-cer,* the other the *Tropick of Capricorn:* they are diſtant from the Equinoctial 23 degrees 30 minutes; and therefore are the bounds of the *Ecliptick.* They receive their names from the Celeſtial Signe that they are joyned unto; as the one the *Tropick of Cancer,* becauſe it touches the Signe of *Cancer;* the other the *Tropick of Capricorn,* becauſe it touches the Signe of *Capricorn.*

IX.

IX. *Of the* Circles *Arctick and* Antarctick.

About the Poles of the World are two fmal Cirdes defcribed; the one called the *Arctick,* the other the *Antartick:* That in the North is called the *Arctick Circle:* that in the South the *Antarctick Circle.*

They have the fame diftance from the Poles of the World that the *Tropicks* have from the Equinoctial Circle, (*viz.* 23 degrees 30 minutes) and that the *Ecliptick* hath from the Poles of the World; and therefore run through the Poles of the *Ecliptick.*

X *Of the* Images *called* Conftellations, *drawn upon the* Celeftial Globe.

Here I think fit to be beholding to Dr. *Hood.* for the paines he hath taken in his comment upon the *Images and Conftelations.* He faith, The ftars are brought into Conftellations, for inftructions fake: things cannot be taught without names, to give a name to every ftar had been troublefome to the Mafter, and for the Scholler; for the Mafter to devife, and for the Scholler to remember: and therefore the Aftronomers have reduced many ftars into one Conftellation, that thereby they may tell the better where to feek them; and being fought, how to exprefs them.

All the Conftellations formerly notified by the Antients were in number 48. twelve whereof we call the twelve Signes of the Zodiack, *viz.* 1 *Aries,* ♈. 2 *Taurus,* ♉. 3 *Gemini,* ♊. 4 *Cancer,* ♋. 5 *Leo,* ♌. 6 *Virgo,* ♍. 7 *Libra,* ♎. 8 *Scorpio,* ♏. 9 *Sagittarius,* ♐. 10 *Capricorn,* ♑. 11 *Aquarius,* ♒. 12 *Pifces.* ♓. One and twenty more are Placed in the North Hemifphear, and are called 1 *Urfa minor,* 2 *Urfa Major,* 3 *Draco,* 4 *Cepheus,* 5 *Bootes,* 6 *Corona Septentri,* 7 *Hercules,* 8 *Lyra,* 9 *Cygnus,* 10 *Caffiopeia,* 11 *Perfeus,* 12 *Auriga,* 13 *Serpentarius,* 14 *Serpens Ophiuchi,* 15 *Sagitta,* 16 *Aquila,* 17 *Delphinus,* 18 *Equiculus,* 19 *Pegafus,* 20 *Andromeda,* 21 *Triangulum.* The other 15 are fcituated in the South Hemifphear, and called 1 *Cetus,* 2 *Orion,* 3 *Eridanus,* 4 *Lepus,* 5 *Canis Major,* 6 *Canicula,* 7 *Argo Navis,* 8 *Hydra,* 9 *Crater,* 10 *Corvus,* 11 *Centaurus,* 12 *Lupus,* 13 *Ara,* 14 *Corona Auftrina,* 15 *Pifces Auftrina.* Befides there are 2 other Conftellations in the North Hemifphear, *viz.* *Antinous,*

D   and

and *Coma Berenices:* which becauſe they were not ſpecified by the Ancients are here inſerted apart.

Now the Aſtronomers did bring them into theſe figures, and not into other, being moved therto by theſe three reaſons: firſt theſe Figures expreſs ſome properties of the ſtars that are in them; as thoſe of the *Ram* to bee hot and dry; *Andromeda* chained, betokeneth impriſonment: the head of *Meduſa* cut off, ſignifieth the loſs of that part: *Orion* with his terrible and threatning geſture, importeth tempeſt, and terrible effects: The *Serpent,* the *Scorpion,* and the *Dragon,* ſignifie poyſon: The *Bull,* inſinuateth a melancholy paſſion: The *Bear* inferreth cruelty, &c. Secondly, the ſtars, (if not preciſely yet after a ſort) do repreſent ſuch a Figure, and therefore that Figure was aſſigned them: as for example, the *Crown,* both North, and South; the *Scorpion* and the *Triangle,* repreſent the Figure which they have. The third cauſe, was the continuance of the memorie of ſome notable men, who either in regard of their ſingular paines taken in Aſtronomy, or in regard of ſome other notable deed, had well deſerved of man kind.

The firſt Author of every particular Conſtellation is uncertain; yet are they of great antiquity; we receive them from *Ptolomie,* and he followed the *Platonicks;* ſo that their antiquity is great. Moreover we may perceive them to be ancient by the Sciptures; and by the Poets. In the 38 Chapter of *Job* there is mention made of the *Pleiades, Orion,* and *Arcturus,* and *Mazzaroth,* which ſome interpret the 12 Signes: *Job* lived in the time of *Abraham,* as *Syderocrates* maketh mention in his Book *de Commenſurandis locorum diſtantiis.*

Now beſides all this, touching the reaſon of the invention of theſe Conſtellations, the Poets had this purpoſe, *viz.* to make men fall in love with Aſtronomy: And to that intent have to every Coſtellation invented ſtrange conceited ſtories, (as you may read at the latter end of this Book) therein imitating *Demoſthenes,* who when he could not get the people of *Athens* to hear him in a matter of great moment, and profitable for the Commom-wealth, he began to tell them a tale of a fellow that ſold an Aſs; by the which tale, he ſo brought on the Athenians, that they were both willing to hear his whole Oration, and to put in practice that whereto he exhorted them. The like intent had the Poets in of thoſe Stories: They ſaw that Aſtronomy being for

com-

commodity fingular in the life of man, was almoft of all men utterly neglected : Hereupon they began to fet forth that Art under Fictions ; that thereby, fuch as could not be perfwaded by commodity, might by the pleafure be induced to take a view of thefe matters : and thereby at length fall in love with them. For commonly you fhall note this, that he that is ready to read the Stories, cannot content himfelf therewith, but defireth alfo to know the Conftellation, or at leaftwife fome principal Star therein.

There are in Heaven yet twelve Conftellations more, pofited about the South Pole, which were added by *Fredericc Houtmanno*, inhabiting on the Ifland *Sumatra* who being accommodated with the Inftruments of that immortal *Tycho*, hath obferved the Longitude and Latitude of thofe Stars, reduced them into Conftellations, and named them as follows, 1 The *Crane*, 2 The *Phenix*, 3 The *Indian*, 4 The *Peacock*, 5 The *Bird of Paradice*, 6 The *Fly*. 7 The *Camelion*. 8 The *South Triangle*, 9 The *Flying Fifh*, 10 *Dorado*, 11 The *Indian Fowl*, 12 The *Southern Serpent*.

XI. *Of the Number of the Stars.*

Although in Heaven there be a very great number of vifible Stars, which for their multitude feem innumerable; yet no wife man will from thence infer that they are impoffible to be counted: for there is no Star in Heaven that may be feen, but its Longitude and Latitude may with meet Inftruments for that purpofe be exactly found; and being once found, it may have a name allotted it, which with its Longitude and Latitude may be Catalogized either for the memory of the Obferver, or the knowledge of Pofterity. Now therefore if any one Star may be obferved, they may all be obferved; and then may they all have Names given them; which tho to the ignorant it feem uncredible, yet to the fons of God, (as *Jofephus* call Aftronomers) who herein participate of their fathers knowledge, it is eafie to *number the Stars, and call them all by their Names Pfal.* 97, 4.

But tho all the Stars in Heaven may be numbred and named, yet have not the Ancient Aftronomers thought fit to take notice of more then 1025 of the chiefeft that are vifible in our Horizon, they being fufficient for any purpofe that we fhall have occafion to apply them unto. Yet of late the induftry of *Frederick Houtman*

D 2   *man*

man aforefaid, hath added to the Catalogue 136 Stars, with their Longitude Latitude and Magnitude, and given Names unto them: which upon my New Globes I have alfo afcerted, as may be feen about the South Pole thereof. So that with thefe 1025, obferved by the Ancients, and thefe 136, the whole number of the Catalogue is 1161.

Some other Stars of late have been alfo obferved by *Bairus,* among the feveral Conftellations aforefaid; but none of any Confiderable Magnitude, and therefore I think fit to pafs them by, and come to their fcituation in Heaven, according to Longitude and Latitude.

XII. *Of the Scituation of the* Stars.

Longitude of the Stars

The Stars are Scituate in Heaven according to their Longitude and Latitude As the Longitude of any Place upon the Terreftrial Globe is an Arch of the Equator, Comprehended between the firft Meridian and the Place. So the Longitude of any Star upon the Celeftial Globe is an Arch of the Ecliptick, contained between the firft point of ♈ and the Star inquired after. But yet becaufe the Ecliptick is divided into twelve Signes, the Longitude of a Star is therefore (in the moft Cuftomary account) an Arch of the Ecliptick comprehended between the Semi-circle of Longitude paffing through the beginning of the Signe the Star is in, and the Semi-circle of Longitude paffing through the Center of the Star.

Latitude of the Stars

The Latitude of a Star is either North or South: North, if on the North fide of the Ecliptick; South, if on the South fide of the Ecliptick. As the Latitude of any Place upon the Terreftrial Globe is an Arch of the Meridian, contained between the Equator and the Parallel of the Place, So is the Latitude of any Star upon the Celeftial Globe an Arch of a Semi-circle of Longitude, comprehended between the Equinoctial and the Star inquired after.

XIII. *Of the* Magnitudes *of the* Stars:

For the better diftinction of the feveral fizes of Stars, they are divided into fix feveral Magnitudes. The biggeft and brighteft Stars are called Stars of the firft Magnitude: Thofe one degree
inferiour

inferiour in light and bigneſs are called Stars of the Second Magnitude, Thoſe again one degree inferiour to the Stars of the ſecond Magnitude, are called Stars of the Third Magnitude, and ſo the Stars gradually decreaſe unto the ſixth Magnitude, which is the ſmaleſt, ſome few obſcure Stars only excepted, which for their Minority and dimneſs are called *Nebula.* Theſe ſeveral Magnitudes of the ſtars are expreſſed on the Globe in ſeveral ſhapes, as may be ſeen in a ſmall Table placed on the Globe for that purpoſe.

Now for your further ſatisfaction and delight, I have inſerted a Collection of D. *Hoods,* wherein is expreſſed the meaſure of The meaſures every Magnitude, and the proportion it hath, firſt, to the Diame- of the ſeveral ter, and ſecondly, to the Body of the Earth. Stars

The greatneſs of any thing (ſaith he) cannot be better expreſſed then by comparing it to ſome common meaſure, whoſe quantity is known : The common meaſure whereby Aſtronomers expreſs the greatneſs of the Stars, is the Earth;

Sometimes they compare them with the Diameter of the Earth, ſometimes with the Globe thereof: The Diameter according to their account which allow but 60 miles to a degree, containeth 6822 $\frac{8}{11}$ miles; and the whole ſoliditie of the Globe containes 165, 042, 481, 283. miles and $\frac{79}{137}$. According to Ptolome, who allotteth to every degree 62 $\frac{1}{2}$ miles, the Diameter containeth 7159 miles $\frac{1}{11}$, and the whole ſoliditie of the Globe, hath 192, 197, 184, 917, $\frac{473}{1331}$ miles.

The proportion of the Diameters of the fixed Stars; Compared with the Diameter of the Earth.

The Diameter of a fixed Star of the firſt Magnitude compared with the Diameter of the Earth hath ſuch proportion to it, as 19 hath to 4 : therefore it containeth the Diameter of the Earth 4 times and $\frac{3}{4}$.

The Diameter of a Star of the ſecond Magnitude is unto the Diameter of the Earth as 269 is to 60 : therefore it containeth it 4 $\frac{29}{60}$ times.

The Diameter of a fixed Star of the third Magnitude is unto the Diameter of the Earth as 25 unto 6 : therefore it containeth it 4 $\frac{1}{6}$ times.

The Diameter of a fixed Star of the fourth Magnitude is unto

D 3   the

the Diameter of the Earth as 19 unto 5 : therefore it containeth it $3\frac{4}{5}$ times.

The Diam. of a fixed Star of the fifth Mag, is unto the Diameter of the Earth, as 119 unto 36. therefore it containeth it $3\frac{11}{36}$ times.

The Diam. of a fixed Star of the ſixth Mag. is unto the Diame- of the Earth, as 21 unto 8 ; therefore it containeth it $\frac{5}{8}$ times.

As for the proportions of the cloudie and obſcure Stars, they are not expreſſed becauſe they are but few, and of no great ac- count in reſpect of their ſmalneſs.

The proportions of the fixed Stars compared with the Globe of the Earth, are as follow.

A Star of the firſt Magnitude is to the Globe of the Earth, as 6859, to 64. therefore it containeth the Globe of the Earth 107 $\frac{1}{6}$ times.

A Star of the ſecond Magnitude is to the Globe of the Earth, as 19465109 is to 216000. therefore it containeth it $90\frac{1}{8}$ times.

A Star of the third Magnitude is to the Globe of the Earth, as 15625 is unto 216 : therefore it containeth it $72\frac{1}{3}$ times.

A Star of the fourth Magnitude is to the Globe of the Earth as 6850 is unto 125 : therefore it containeth the Globe of the Earth $54\frac{11}{12}$ times.

A Star of the fifth Magnitude is to the Globe of the Earth, as 1685159 : is unto 46656 : therefore it containeth the Globe of the Earth $36\frac{1}{8}$ times.

A Star of the ſixth Magnitude is to the Globe of the Earth, as 9261 is unto 512 : therefore it containeth the Globe of the Earth $18\frac{1}{10}$ times.

I confeſs all this may ſeem matter of incredulity to thoſe whoſe underſtanding is ſwayed by their viſual ſence; but if they be ca- pable to conſider the vaſt diſtance of thoſe Huge Bodies, (the Stars) from the face of the Earth, and alſo the diminutive qua- lity of Diſtance, their reaſon will be rectified, and their increduli- ty turn'd into an acknowledgement of the unſpeakable wiſdom of Almighty God; and they will ſay with the *Pſalmiſt, Great is our Lord, Great is his Power, his Wiſdom is infinite. Pſal.* 147. 5.

The diſtance of the Stars therefore from the Earth, is accor- ding to M. *John Dee's* Computation, $20081\frac{1}{2}$ Semidiameters of the Earth. The Semidiameter of the Earth containeth of our

com-

common miles 3436 $\frac{4}{11}$, Such miles as the whole Earth and Sea round about is 21600 : allowing for every degree of the greateſt Circle 60 miles : ſo that the diſtance of the Stars from the Earth is in miles 69006540. Now as M. *Dee* ſaith, (almoſt in theſe ſame words) if you weigh well with your ſelf this little parcel of fruit Aſtronomical; as concerning the bigneſs and diſtance of the Stars, &c. and the Huge maſſineſs of the Starry Heaven, you will find your Conſciences moved with the Kingly Prophet to ſing the confeſſion of Gods Glory; and ſay, *The Heavens Declare the Glory of God, and the firmament ſheweth forth the works of his Hands.*

XIIII. *Of the* Nature *of the* Stars.

To many of the Principal Stars there is in Planetical Charaĉters prefixed their Planetical Natures. The Aſtrologers make great uſe of them for knowing the nature of the Stars : for thoſe Stars that have the charaĉter of ♄ adjoined are ſaid to be of the nature of ♄: thoſe that have ♃ adjoined, are of the nature of ♃: and ſo of the reſt. If a Star have the charaĉters of two Planets adjoined, that Star participates of both their Natures, but moſt of that Planets whoſe charaĉter is firſt placed.

The uſe Aſtronomers make of thoſe charaĉters, is for knowing that culler of any Star; as if a Star have ♄ adjoined, it is of the culler of ♄; if ♃, it is the culler of ♃, &c.

The fixed Stars are known from the Planets by their continual twinckling; for the Planets never twinckle, but the fixed Stars do.

XV. *Of Via Laĉtea, or the* Milky way.

This ſubjeĉt becauſe it is already ſo fully handled by Dr. *Hood,* that more then he hath written cannot well be ſaid, either of his own oppinion or other mens, I think fit therefore to give you his own words: which are as follow.

V I A L A C T E A. or *Circulus Laĉteus;* by the Latines ſo called; and by the Greekes, *Galaxia;* and by the Engliſh, the *Milky way.* It is a broad white Circle that is ſeen in the Heaven: In the North Hemiſphear, it beginneth at *Cancer,* on each ſide the head thereof, and paſſeth by *Auriga,* by *Perſeus,* and

and *Caſſiopeia,* the *Swan,* and the head of *Capricorn,* the tayl of
Scorpio, and the feet of *Centaur, Argo* the *Ship,* and ſo unto the
head of *Cancer.* Some in a ſporting manner, do call it *Watling
ſtreet;* but why they call it ſo, I cannot tell; except it be in regard of the narrowneſs that it ſeemeth to have, or elſe in reſpect
of that great High way that lieth between *Dover* and *S. Albons,*
which is called by our men *Watling ſtreet.*

Concerning this Cirde there are ſundry opinions : for there is
great difference among ſome writers, both touching the place,
matter, and efficient cauſe thereof. *Ariſtotle* diſſenteth from
all other, both Philoſophers and Poets, in the place, matter, and
cauſe of this Circle; ſaying, that it is a Meteor ingendred in the
Air, made of the vapors of the earth, drawn up thither by the
heat of the Sun, and there ſet on fire. But his opinion is of all men
confuted.

Firſt, touching the place, it cannot be in the Air; for whatſoever is in the Air, is not ſeen of all men, at all times, to be under one and the ſame part of Heaven. If we ſee it in the South,
they that are in the Weſt ſhall ſee it under the Eaſt ſide of the
Heaven; and they that are in the Eaſt, ſhall ſee it in the Weſt
part of the Heaven; but this Cirde is of all men ſeen alwaies under the ſame part of Heaven, and to be joyned with the ſame
Stars; therefore it cannot be in the Air.

Again, for the matter, it cannot be made of that which *A-
riſtotle* nameth (*i. e.*) the vapours of the earth, becauſe of the
long continuance of the thing, and that without any alteration :
for it is impoſſible that any Meteor made of vapours drawn up
from the water, or exhalations from the earth, ſhould laſt ſo
long; as may be ſeen in blazing Stars; which though they have
continued long, as namely, 16. moneths, ſome more, ſome leſs;
yet at the length they have vaniſhed away : whereas this Cirde hath continued from the beginning unto this day. Beſides,
put caſe it were made of theſe exhalations, Whence will they infer the uniformity thereof? The Comets do alter diverſly, both
in the faſhion of their blazing, and alſo in their ſeveral quantities;
whereas in this Cirde, there is nothing but the ſame part, alwaies
of one form and of one bigneſs. In the effcient cauſe therefore
he muſt needs err : for if it be neither in the Air, nor made of the
exhalations of the earth, it cannot be cauſed by the Sun; for the
                                                                one

one is the place and the other the matter, wherein, and whereupon the Sun sheweth his power.

All other, (besides *Aristotle*) agree in the place, but differ in the efficient cause thereof : and they are either Philosophers, or Poets. Both these affirm that it is in the Firmament (*i. e.*) in the eight Sphear; but they disagree in the cause thereof.

The Philosophers (and chiefely *Demecritus*) affirm the cause of the thing, to be the exceeding great number of Stars in that part of Heaven, whose beams meeting together so confusedly, and not coming distinctly to the ey, causeth us to imagine such a whiteness as is seen. But the best opinion is this, that this *Milky way* is a part of the Firmament, neither so thin as the other parts thereof are, not yet so thick as the Stars themselves. If it were as thin as the other parts of the Heaven besides the Stars, then could it not retain the light, but the light would pass through it and not be seen : If it were as thick as the Stars, then would the light be so doubled in it, that it would glister and shine, as the Stars themselves do : but being neither so thin as the one, nor so thick as the other, it becommeth of that whiteness we see.

Blaeu saith, This *Lactean* whiteness and clearness ariseth from a great number of little Stars, constipated in that part of Heaven: flying so swiftly from the sight of our eyes, that we can perceive nothing but a confused light : this the *Tubus Diopticus* (more lately found out) doth evidently demonstrate to us : by the benefit of which little Stars (otherwise inconspicuous to our eyes) are there clearly discerned.

About the Southern Pole are seen two white spots, like little clouds, colured like the *via Lactea*. One of which is trebble the Latitude of the other; some Mariners call them *Nubecula Magellani*.

This *Milkie way* is discribed on the Globe between two tracks of smal Pricks, running through the Images mentioned in the beginning of this Section.

Thus have you the definition of the Globes; with the description of all the lines, Circles, &c. described thereon. I shall now explain unto you the meaning of Several words of Art, which in the use of them you will meet with, and then come to the Use it self.

And first, what is meant by the word *Horizon.*

E   When

When I ſpake of the *Horizon* before, I only mentioned the wooden *Horizon* or frame about the Globe; which becauſe it repreſents the *Mathematical Horizon,* is therefore called the *Horizon:* but the word *Horizon* is to be conſidered more particularly, two manner of waies: as

Firſt, the *Natural Horizon.*

Secondly, the *Mathematical Horizon.*

The *Natural Horizon* is that Appearent Circle which divides the Viſible part of Heaven from the inviſſible; it extends it ſelf in a ſtraight line from the Superficies of the Earth, every way round about the place you ſtand upon, even into the very Circumference of the Heavens. It is onely diſcerned at Sea, or on plaine ground, that is free from all hinderances of the ſight as Hills, Trees, Houſes, &c.

The *Mathematical Horizon* (which indeed is meant in this Treatiſe, ſo oft as I ſhall have occaſion to name the word *Horizon*) is a great Circle which divides that part of Heaven which is above us, from that which is under us, preciſely into two equal parts: whoſe *Poles* are the *Zenith* and *Nadir.* In this Circle the *Azimuths* or *Verticle Circles* are numbred: and by this Cirde our Daies and Nights are meaſured out unto us: for while the Sun is above the *Horizon* it is day; and when it is under the *Horizon* it is Night.

This Circle is repreſented unto us by the upper Plain of the wooden *Horizon* : Therefore ſo oft as you are directed to bring any degree or Star &c. to the *Horizon,* it muſt be underſtood that you muſt turn the Globe till the degree or Star come juſt to the upper inner edge of the wooden *Horizon.*

Zenith.
Nadir.

The *Zenith,* and *Nadir,* are two points oppoſite to one another. The *Zenith* is that point in Heaven which is directly over our Heads : and the *Nadir* is that point in Heaven which is directly under our feet.

Azimuths, or *Verticle Circles.*

The *Azimuths* or *Verticle Circles* are great Circles paſſing through the *Zenith,* and *Nadir,* whoſe *Poles* are the *Zenith* and *Nadir.* And as the *Meridians* cut the *Equator,* and all *Parallels* to the *Equator* at Right Angles, ſo the *Azimuths* cut the *Horizon* and all *Almicanthars* at Right Angles alſo. The *Azimuths* (as the *Meridians*) are infinite; and are numbred by degrees from the Eaſt and Weſt point towards the North and South in the Horizon : as alſo is the *Amplitude.*

The

The *Almicanthars* are Circles Parallel to the Horizon, whose Poles are the *Zenith* and *Nadir.* They are also called *Circles of Altitude,* because when the Sun Moon or any Star, is in any number of degrees above the *Horizon,* it is said to have so many degrees of *Altitude,* which degrees of *Altitude* are numbred upon the *Verticle Circle* from the *Horizon* upwards, towards the *Zenith.* The *Almicanthars* are also infinite : as *Parallels, Meridians* and *Azimuths* are. — *Almicanthars, or Circles of Altitude.*

The *Amplitude* is the number of degrees contained between the true East or West point in the *Horizon,* and the rising or setting of the Sun, Moon, or Stars. &c. — *Amplitude.*

The *Declination* is the number of degrees that the Sun, Moon, or any Star, is distant from the *Equinoctial,* towards either *Pole :* and hath a double Denomination, *viz.* *North Declination,* and *South Declination :* for if the Sun Moon or Star swarve towards the North *Pole,* they are said to have *North Declination;* if towards the South Pole, *South Declination.* — *Declination.*

The *Right Ascension* is the number of degrees of the *Equinoctial* (accounted from the first point of *Aries*) which comes to the *Meridian* with the Sun Moon or Star, or any other point in *Heaven* proposed. — *Right Ascension.*

The *Oblique Ascension* is the number of degrees of the *Equinoctial* which comes to the East side of the *Horizon* with the Sun Moon or any Star. — *Oblique Ascension.*

The *Oblique Descension* is the degrees of the *Equinoctial* which comes to the West side of the *Horizon* with the Sun Moon or any Star. — *Oblique Descension.*

The *Ascensional Difference* is the number of degrees after subtraction of the *Oblique Ascension* from the *Oblique Descension,* or *Oblique Descension* from the *Oblique Ascension.* — *Ascensional Difference.*

So many degrees as you are said to sail towards the *Pole,* you are said to *Raise the Pole;* and so many degrees as you sail from the *Pole,* you are said to *Depress the Pole.* — *Raise the Pole. Depress the Pole.*

Course, is the point of the Compass you sail upon; as if you sail *East-wards,* it is an *Easterly Course,* if *West,* a *Westerly Course* &c. — *Course.*

Distance is the number of leagues you have sailed from any Place, upon any *Course.* — *Distance.*

A *Zone* is a space of Earth contained between two Parrallels. The ancient *Geographers* made five *Zones* in the *Earth.* Two *Frozen,* Two *Temperate,* and one *Burnt Zone.* — *Zone.*

The

Frozen Zones.

The two *Frozen Zones* are thoſe parts of the *Globe*, comprehended between the North Pole and the *Arctick Circle*, and the South Pole and the *Antarctick Circle*; by the Ancients called inhabitable; becauſe the Sun being alwaies far remote from them, ſhoots its beams *Obliquely* upon them, which *Oblique* beams are ſo very weak, that all their Summer is but a continued Winter, and the Winter (as they thought) impoſſible to be at all indured.

Temperate Zones.

The *Temperate Zones* are the ſpace of *Earth* contained between the *Arctick Circle* and the *Tropick* of ♋, and the *Antarctick Circle* and the *Tropick* of ♑: by the Ancients called *Temperate* and Habitable; becauſe they are compoſed of a ſweet Mediocrity, between outragious Heat and extremity of Cold.

Burnt Zone.

The *Burnt Zone* is the ſpace of *Earth* contained between the *Tropick* of ♋, and the *Tropick* of ♑, called by the Ancients *Unhabitable*; becauſe in regard the Sun never moves out of this *Zone* but darts its Beames perpendicularly upon it, they imagined the *Air* was ſo unſufferable *Hot*, that it was impoſſible for any to inhabite in this *Zone*. So that as you ſee they held the two *Temperate Zones* only habitable; and the two *Frozen Zones* and one *Burnt Zone*, altogether unpoſſible to be inhabited. But their Succeſſors either animated by induſtry, or compeld by neceſſity, have apparently confuted that Aſſertion; for at this time many thouſands can witneſs that their bloods are not ſo greaſie as to be melted in the Scortching heat of the one, or ſo watry as to be congealed in the Icy froſts of the other.

Climates.

The Ancients have yet otherwiſe divided the *Earth* into four and twenty *Northern Climates*, and four and twenty *Southern Climates*: ſo that in all there is eight and forty *Climates*. The *Climates* are altered according to the half hourly increaſing of the longeſt daies; for in the Latitude where the longeſt daies are increaſed half an hour longer then they are at the Equator (*viz.* longer then 12 hours) the firſt *Climate* begins ; and in the Latitude where they are increaſed an whole hour longer then in the Equator, the ſecond *Climate* begins; where the daies are increaſed three half hours longer then in the Equator, the third *Climate* begins; and ſo onwards, the *Climates* alter according as the longeſt day increaſes half an hour, till you come to find the longeſt day 24 hours long

Now the Ancients (in thoſe times) knowing no more then nine Habitable *Climates*, gave names only to nine. The firſt they

they called *Dia Meroes*, after the name of a famous *Inland Iland*, which is fcituate about the middle of that *Climate*, and is now called *Gueguere*. The fecond *Climate* they called *Dia Syenes*, after the name of an eminent *Citty in Egypt*, lying about the midft of that *Climate*. The third *Dia Alexanderas*, after the name of the *Metropolitan Citty* of *Egypt*. The fourth *Dia Rhodes*. The fifth *Dia Romes*. The fixth *Dia Ponton*. The feventh *Dia Boriftheneos*, The eighth *Dia Ripheos*. The ninth *Dia Daniam*.

Thefe names belong only to the *Climates* on the North fide of the Equator. But thofe on the South fide (in regard of the fmal Difcoveries thofe Ages had on that fide the Equator) were diftinguifht only by the addition of the word *Anti*, to the fame *Southerly Climate* : as the firft *Southern Climate* (which is that *Climate* that lies as many degrees to the South-ward as the firft doth to the North-ward) they called *Anti Meroes*. The fecond *Anti Syenes*. The third *Anti Alexanderas* : and fo on to the ninth.

In every *Climate* is included two Parallels, which are of the *Parallels.* fame nature with the *Climates*, fave only that as the *Climates* alter by the half hourly increafing of the longeft day, the *Parallels* alter by the quarter hourly increafing of the longeft day.

Furthermore, in refpeét of the *Horizon*, we find the *Sphear* conftituted into a threefold Pofition: as firft, into a *Direét Sphear,* Secondly, a *Parallel Sphear,* Thirdly, an *Oblique Sphear.*

A *Direét Sphear* hath both the Poles of the World in the *Direét Sphear.* *Horizon,* and the *Equinoétial* tranfiting the *Zenith*. In a *Direét Sphear* all the *Circles Parallel* to the Equator make right angles with the *Horizon*, and are alfo divided into two equal parts by the *Horizon* : and in a *Direét Sphear* the Sun Moon and Stars are alwaies twelve hours above the *Horizon*, and twelve hours under the *Horizon*, and confequently make twelve hours *Day,* and twelve hours *Night*.

It is called a *Direét Sphear* becaufe all the *Celeftial Bodies,* as Sun Moon and Stars &c. by the *Diurnal Motion* of the *Primum Mobile*, afcend direétly above, and defcend direétly below the *Horizon*.

They that inhabite under the Equator have the *Sphear* thus pofited; as in the *Iland Borneo, Sumaira, Celebes, St. Thomas* a great part of *Africk, Peru* in the *Weft-Indies* : &c. as you may

E 3   fee

ſee by the *Globe* it ſelf; if you move the *Braſen Meridian* through the notch in the *Horizon,* till the Poles thereof touch the *Horizon.* As in this Figure.

Parallel Sphear.

    A *Parallel Sphear* hath one Pole of the VVorld in the *Zenith,* the other in the *Nadir,* and the *Equinoctial* line in the *Horizon.*

    In a *Parallel Sphear* all the *Circles Parallel* to the *Equinoctial* are alſo *Parallel* to the *Horizon,* and in a *Parallel Sphear* from the 10th of *March* to the 11th of *September* (the Sun being then in the *Northerly Signes* and conſequently on the North ſide the *Horizon*) there is ſix Moneths Day in the North, and ſix Moneths Night in the South : and contrarily from the 11th of *September* to the 10th of *March,* (the Sun being then in the *Southerly Signes,* and therefore on the South ſide the *Horizon*) there

there is fix Moneths Day in the South, and fix Moneths Night in the North.

It is called a *Parallel Sphear*, becaufe the Sun Moon or Stars in a *Diurnal Revolution* of the *Heavens*, neither afcend higher or defcend lower, but alwaies move *Parallel* to the *Horizon*.

The *Earth* is thus Pofited under both the Poles, *viz.* in 90 degrees of Latitude; as may be feen by the *Globe*, if you turn the *Brafen Meridian* till either of the Poles be elevated 90 degrees above the *Horizon*. As in this figure.

An *Oblique Sphear* hath the *Axis of the World* neither *Di-* *Oblique* rect nor *Parallel* to the *Horizon*, but lies aflope from it.   *Sphear.*

In an *Oblique Sphear* all the *Celeftial Bodies*, as Sun Moon or Stars &c. have (in refpect of the *Horizon*) Oblique and un-equal Afcenfions and Defcenfions, and all the lines Parallel to the Equator

Equator make unequal Angles with the *Horizon,* and are cut by
the *Horizon* into unequal parts ; for thofe lines towards the
elevated Pole, have a greater portion of a Circle under the *Hori-*
zon then above it: only the Equator becaufe it hath the fame
Center with the *Horizon,* doth divide the *Horizon* into two e-
qual parts, and is alfo divided into two equal parts by the
Horizon.

Hence is follows that when the Sun is in any part of the *E-*
cliptick that declines towards the elevated Pole, the Daies in the
elevated *Hemifphear* fhall be longer then the Nights : and when
the Sun is in any part of the *Ecliptick* that declines towards the
Depreffed Pole, the Nights fhall be longer then the Daies. But
when the Sun is in the *Equinoctial,* (becaufe whether the Pole be
either Raifed or Depreffed) equal portions remain both above
and under the *Horizon,* therefore the Daies are of the fame
length with the Nights, and the Nights with the Daies.

Alfo in an *Oblique Sphear,* all thofe Stars that have as great or
greater number of degrees of Declination then is the elevated
Poles Complement of Latitude to 90, never fet or come under
the *Horizon,* and thofe Stars that have the fame Declination a-
bout the Depreffed Pole never rife.

It is called an *Oblique Sphear,* becaufe all the Circles of the
Sphear move *Obliquely* about the *Horizon.*

The *Earth* is thus *Obliquely* pofited to all thofe *Nations* that
inhabite under any degree of Latitude either North or Southwards between the Equator and either Pole: as may varioufly be
feen by the Globe, when the Axis lies not on the *Horizon,* nor
the Equator is Parallel to the *Horizon.*   As in this following
Figure.

Moreover all Places have their *Antipodes,* Peræci and
Antæi.

The *Antipodes* of any Place is the oppofite degree on the *Antipodes.*
Globe. As if a Perpendicular were let fall from the Place you
ſtand on, through the Center of the Earth, and continued till it
paſs quite through the Superficies of the *Earth,* on the other ſide;
then in the point where the Perpendicular cuts the Superficies
of the Earth on the other ſide, is the *Antipodes* of that Place.

The Inhabitants of any two Places that are in *Antipodes* to
each other, go with their Feet directly againſt one another : and
have a contrariety in the Seaſons of the Year, and Riſings, and
Settings, of the Sun Moon Stars, and all other of the *Heavenly*
Bodies : ſo that when with us it is Spring, with them it is Autumn; when with us the Sun Riſes, in our *Antipodes* it Sets; and
therefore their Morning is our Evening, their Noon our Mid-
F   night,

night, their Evening our Morning; and their Longeſt Day our ſhorteſt.

Periæci.

The *Periæci* of any Place is that point in the ſame Parallel which comes to the Meridian with the *Antipodes*.

In the *Periæci* of any Place, there happens not that Contrariety of Seaſons in the Year, that doth in the *Antipodes* ; nor in the Length of Daies : for the Daies in both Places are of equal length: but in the times of the Day, there is the ſame contrariety, for (though their Spring be our Spring, and thereſt of their Seaſons of the year the ſame with ours, yet) their Morning is our Evening, their Night our Day, &c.

Antæci

The *Antæci* of any Place is the point under the ſame Meridian that is diſtant from the Equator on the South ſide ſo many degrees as your Place is diſtant from the Equator on the North ſide.

In the *Antæci* there happens not that contrariety in the Daies as doth in the *Antipodes*, but in the Seaſons of the Year there is the ſame contrariety; for in our *Antæci* their Morning is our Morning, their Noon our Noon, their Night our Night: but herein is the Difference, their Spring is our Fall, their Summer our VVinter, &c. and their Longeſt Day our ſhorteſt: as in the *Antipodes*.

The Second BOOK.

Shewing the Practical Use of the

GLOBES.

Applying them to the Solution of Aſtronomical *and* Geographical *Problemes.*

P R Æ F A C E.

Some Advertiſements in Chooſing and Uſing
the GLOBES.

I.

 EE the *Papers* be well and neat-
ly *paſted* on the *Globes* : which
you may *know,* if the *Lines* and
Circles diſcribed thereon meet ex-
actly, and continue all the way
even and whole : the lines not
ſwerving out or in , and the *Cir-*
cles not breaking into ſeveral *Ar-*
ches; nor the *Papers* either come
ſhort, or lap over one the other.

2. See that the *Culler* be tranſparent, and ly not too thick
on the *Globe;* leſt it hide the ſuperficial *Deſcriptions.*

3. See the *Globe* hang evenly between the *Meridian* and
Horizon, not inclining more to one ſide then the other.

4. See the *Globe* ſwim as cloſe to the *Meridian* and *Ho-*
F 2   *rizon*

rizon as conveniently it may ; *leſt you be too much puzzeld
to find againſt what point of the Globe any degree of the Horizon or Meridian is.*

5. *See the Equinoctal line be one with the Horizon,
when the Globe is ſet in a* Parallel Sphear.

6. *See the Equinoctal line cut the Eaſt and Weſt point of
the Horizon, when the Globe is ſet to an* Oblique Sphear.

7. *See the Degrees marked with* 90. *and* 00, *hang exactly over the Equinoctial line of the Globe.*

8. *See that exactly half the Meridian be above the Horizon, and half under the Horizon :* which you may know if
you bring any of the Decimal Diviſions to the North Side
of the Horizon, and find their Complement to 90. in the
South.

9. *See that when the Quadrant of Altitude is placed at
the Zenith, the Beginning of the Graduations reach juſt to
the ſuperficies of the Horizon.*

10. *See that while the Index of the Hour Circle (by the
motion of the Globe) paſſes from one hour to the other,* 15.
degrees of the Equator paſs through the Meridian.

11. *If you have a Circle of Poſition, ſee the Graduations
agree with thoſe of the Horizon.*

12. *See that your wooden Horizons be made ſubſtantial
and ſtrong ;* for (beſides the Inconveniences that thin wood
is ſubject unto, in reſpect of warping and ſhrinking) I have
had few Globes come to mending that have not had either
broken Horizons, or ſome other notorious fault, occaſioned
through the ſleightneſs of the Horizons.

In the Uſing the Globes.

KEep *the Eaſt ſide of the Horizon alwaies towards you,
unleſs your Propoſition requires the turning of it :*
which Eaſt ſide you may know by the Word Eaſt, placed on
the outmoſt verge thereof. For then have you the graduated

ated *fide of the* Meridian *alwaies towards you ; the* Quadrant *of altitude before you, and the* Globe *divided exactly into two equal parts.*

So oft as I name to, at, of, *or* under the Meridian, *or* Horizon, *I mean the* East *fide of the* Meridian, *and Superficies of the* Horizon *: becaufe the* East *fide of the* Meridian *paffes through the* North *and* South *points, both of the* Globe *and* Horizon; *and agrees just with the middle of the* Axis *:* And *the Superficies of the* Horizon *divideth the* Globe *exactly into two equal parts.*

It you happen to ufe the Globes *on the* South *fide the* Equator, *you must draw the* wyers *out of either* Pole, *and change them to the contrary* Poles *; putting the longest wyer into the* South Pole. And *becaufe on the other fide the* Equator *the* South Pole *is elevated, therefore you must elevate the* South Pole *of the* Globe *above the* Horizon *; according to the* South Latitude *of your* Place *; as fhall be fhewed hereafter.*

In the working fome Problemes *it will be required that you turn the* Globe *to look on the* West *fide thereof : which turning will be apt to jog the* Ball, *fo as the degree that was at the* Horizon *or* Meridian, *will be moved away, and thereby the* Pofition *of the* Globe *altered. To avoid which inconvenince you may make ufe of a* Quill, *thrusting the* Feather *end between the* Ball *and the* Brazen Meridian, *and fo wedge it up, without wronging the* Globe *at all, till your* Propofition *be anfwered.*

PROBLEME I.

To *find the* Longitude *and* Latitude *of* Places, *on the* Terrestrial Globe.

S Eek the Place on the Terreſtrial Globe, whoſe Longitude and Latitude you would know, and bring that Place to the

Brazen Meridian ; and fee how many degrees of the Equator is cut by the Meridian, from the firft general Meridian, (which on my Globes pafs through *Gratiofa,* one of the Ifles of the *Azores,*) for that number of degrees is the Longitude of the Place.

Example.

I defire to know the Longitude of *London,* and clofe to the name *London* I find a fmal mark o thus, (which fmal mark is in fome Globes and Maps adorned with the Picture of a Stee-ple, &c.) therefore I do not bring the word *London* to the Me-ridian, but that fmal mark ; for that alwaies reprefents the the Town or Citty fought for : And keeping the Globe fteddy in this Pofition, I examine how many degrees of the Equa-tor are contained between the Brazen Meridian, and the firft ge-neral Meridian; which I find to be 24. deg. oo. min. There-fore I fay the Longitude of *London* is 24. degrees oo. min.

For the Latitude.

See on the Brazen Meridian how many degrees are contained between the Equator and the mark for *London* ; which in this Example is $51\frac{1}{2}$: therefore I fay *London* hath $51\frac{1}{2}$ degrees North Latitude.

PROBLEME II.

The Longitude *and* Latitude *being known, to* Rectifie *the* Globe *fit for ufe.*

1. WHen you rectifie the Globe to any particular Lati-tude you muft move the Brazen Meridian through the notches of the Horizon till the fame number of degrees accounted on the Meridian from the Pole (about which the Hour-Circle is) towards the North point in the Horizon (if in North Latitude, and toward the South if in South Latitude) come juft to the edge of the Horizon.

Example.

By the former Propofition I found the Latitude of *London* to be
$51\frac{1}{2}$ de-

51½ degrees North Latitude : therefore I count 51½ degrees from the Pole downwards towards my right hand, and turn the Meridian through the notches of the Horizon till thofe 51½ degrees comes exactly to the uppermoft edge of the North point in the Horizon; and then is the Meridian rectified to the Latitude of *London*.

2. Next rectifie the Quadrant of altitude, after this manner,

Screw the edge of the Nut that is even with the graduated edge of the thin Plate, to 51½ degrees of the Brazen Meridian, accounted from the Equinoctial on the Southern fide the Horizon, which is juft the *Zenith* of *London* : and then is your Quadrant Rectified.

3. Bring the degree of the Ecliptick the Sun is in that day, to the Meridian : which you fhall learn to know by the next Probleme, and then turn the Index of the Hour Circle to the hour 12. on the South fide the Hour Circle, and then is your Hour Circle alfo rectified fit to ufe, for that Day.

4. Laftly If you will rectifie the Globe to correfpond in all refpects with the Pofition and Scituation of the Sphear, you muft fet the four Quarters of the Horizon. viz. *Eaft, Weft, North,* and *South,* agreeable with the four quarters of the World ; which you may do by the Needle in the bottom of the Horizon ; for you muft turn the Globe fo long till the Needle point juft to the Flower de luce. Next you muft fet the Plain of the wooden Horizon parallel to the Horizon of the World ; which you may try by fetting a common Level on the four Quaters of the Horizon. And then pofiting the degree of the Ecliptick the Sun is in, to the Height above, or depth below the Horizon, the Sun hath in Heaven, (as by the 11th Probleme) your Globe is made Correfpondent in all points with the frame of the Sphear, for that particular Time, and Latitude.

PROBLEME III.

To find the Place of the Sun in the Ecliptick, the Day of the Moneth being firft known.

S Eek the Day of the Moneth in the Circle of Moneths upon the Horizon, and right againft it in the Circle of Signes is the degree of the Ecliptick the Sun is in.

Exam.

Example.

Imagine the Day to be given is *May* 10. therefore I ſeek on the Horizon in the Circle of Moneths, for *May*, and find the Moneths divided into ſo many parts as there is Daies in the Moneth; which parts are marked with Arithmetical figures, from the beginning of the Moneth to the end, and denote the number of the Day of the Moneth that each Diviſion repreſents : therefore among the Diviſions I ſeek for 10, and directly againſt it in the Circle of Signes, I find ♉ 29. degrees. Therefore I ſay *May* 10. the Suns Place is in 29. degrees of ♉.

But note, that if it be *Leap Year*, inſtead of the 10. of *May* you muſt take the 11. of *May* : becauſe *February* having in a *Leap Year* 29. Daies, the 29. of *February* muſt be reckoned for the firſt of *March*, and the firſt of *March* for the ſecond of *March* ; the ſecond of *March* for the third of *March* ; and ſo throughout the year.

The *Leap Year* is cauſed by the ſix od hours more then 365. daies that are aſſigned to every common Year : ſo that in a Revolution of 4. Years, one Day is gained, which is added to *February* ; and therefore *February* hath every fourth or *Leap Year* 29. Daies.

PROBLEME IIII.

To find the Day of the Moneth, the Place of the Sun being given.

AS in the laſt Probleme it was your task to find on the Horizon the Day of the Moneth firſt, ſo now you muſt firſt ſeek the Signe and degree the Sun is in, and againſt it in the Circle of Moneths you ſhall ſee the Day of the Moneth : As againſt ♉ 29. you have May 10.

PROBLEME V.

The Place of the Sun given, to find its Declination.

HAving by the third Probleme found the Suns Place on the Plain of the Horizon, you muſt ſeek the ſame degree in
the

the *Ecliptick* on the Globe ; then bring that degree to the Brazen Meridian; and the number of degrees intercepted between the Equinoctial and the degree juft over the degree of the Ecliptick the Sun is in, is the Declination of the Sun for that Day : and bears its Denomination of North or South, according to its Pofition either on the North or South fide the Equinoctial.

Example.

By the third Probleme aforefaid, of *May* 10. I find ♉ 29. the Suns Place; Therefore I feek in the *Ecliptick* Line on the Globe for ♉ 29. and bring it to the Eaft fide of the Brazen Meridian, which is the graduated fide; and over ♉ 29. I find on the Brazen Meridian 20. deg. 5. min. (numbred from the Equinoctial:) and becaufe ♉ is on the North fide the Equinoctial, therefore I fay, The Sun hath *May* 10. North Declination 20. degrees 5. min.

PROBLEME VI.

The Place of the Sun given, to find its Meridian Altitude.

THe Globe rectified, Bring the degree of the Sun to the Meridian, (or which is all one, the degree of the *Ecliptick* the Sun is in;) and the number of degrees contained between the Horizon and the Suns Place in the Meridian, is the number of degrees that the Sun is Elevated above the Horizon at Noon, or (which is all one) the Meridian Altitude of the Sun.

Example.

To know what Meridian Altitude the Sun hath here at *London*, *May* 10. I bring the Suns Place (found by the third Probleme) to the Meridian, and count on the Meridian the number of degrees contained between the Horizon and the degree juft over the Suns Place ; which in this Example I find to be 58 $\frac{1}{2}$. Therefore I fay the Suns Meridian Altitude *May* 10. is here at *London* 58 $\frac{1}{2}$ degrees.

G   PROB.

PROBL. VII.

*The Suns Place given, to find the Hour of Sun Riſing,
and the length of the Night and Day.*

THe Globe and Hour Index rectified, Seek the degree the
Sun is in on the Globe, and bring that degree to the Eaſtern Side of the Horizon ; and the Index of the Hour Circle
will point at the Hour of Sun Riſing.

Example.

To know the Hour of Sun Riſing here at *London, May* 10.
The Suns Place (as before) is ♉ 29. Therefore the Globe
being rectified (as before) I ſeek ♉ 29. degrees on the Globe,
and bring that degree to the Eaſt Side of the Horizon; and looking on the Index of the Hour Circle, I find it point at 4. a
clock and $\frac{1}{6}$ part of an hour more towards 5 ; therefore I ſay
May 10. the Sun riſes here at *London* at $\frac{1}{6}$ (which is 12. minutes) after 4 a clock in the Morning.

If you double 4 hours 12. minutes, it gives you the length of
the Night, 8 hours 24. minutes. And if you ſubſtract the
length of the Night 8. hours 24. minutes, from 24. hours, the
length of Day and Night; it leaves the length of the Day 15.
hours 36. minutes.

PROB. VIII.

To find the Hour of Sun Set.

TUrn the Place of the Sun to the Weſt ſide of the Horizon,
and the Index of the Hour Circle ſhews on the Hour-Circle the hour of Sun ſet ; which on the 10th of *May* aforeſaid,
is $\frac{4}{6}$ parts of an hour after 7. a clock at Night, *Viz.* the Sun
Sets at 48. minutes paſt 7. a clock.

PROB.

PROB. IX.

To find how long it is Twilight in the Morning, and Evening.

Wilight is that promiſcuous and doubtfull light which appears before the Riſing of the Sun in the Morning, and continues after the ſetting of the Sun in the Evening: It is made by the extenſion of the Suns beams into the Vapours of the Air, when the Sun is leſs then 18. deg. below the Horizon: for the Sun ere it Riſes, and after it Sets, ſhoots forth its Beams through the Air, and ſo illuminates the Vapours of the Air ; which illumination does by degrees enlighten the Horizon, and ſpreads through the *Zenith*, even into the Weſt, ere the Sun Riſes ; and alſo continues above the Horizon, after the Sun ſets.

Now though it be Twilight when the Sun is 18. degrees below the Horizon; yet the duration of Twilight (is alterable both in reſpect of *Time*) and *Place*: for at ſuch Time at the Sun is fartheſt diſtant from any Place, the Twilight ſhall be greater, then when it is neereſt. And in reſpect of *Place*, All Places that have great Latitude from the Equator, have longer Twilight than thoſe that are neerer to the Equator : for as Authors ſay, under the Equator there is no Twilight; when again in many Climes both Northward and Southward, the Nights are indeed no Nights but only (as it were) a little over-ſpread with a cloudy Shade ; and is either increaſed or diminiſhed according to the cautation of Meoterological Cauſes.

Therefore to know the beginning of Twilight in the Morning here at *London*, *May* 10; you muſt (having the Globe rectified) turn the degree of the *Ecliptick* which is oppoſite to the Place of the Sun till it be elevated 18. degrees in the Quadrant of Altitude above the Horizon in the *Weſt* ; So ſhall the Index of the Hour-Circle point at the Hour that Twilight begins: Then ſubtract the Hour and Minute that Twilight begins from the Hour and Minute of Sun Riſing, if in the Morning, or ſubſtract the Hour of Sun ſett from the Hour of Twilight, if at Night ; and the remainder is the length of Twilight.

Example.

The Globe Quadrant and Hour-Index being rectified, as before;

fore; and the Suns place given, ♉ 29. I ſeek the oppoſite degree on
the Globe, after this manner· I bring ♉ 29. to the Meridian, and
obſerve what degree of the *Ecliptik* the oppoſite part of the
Meridian cuts ; and becauſe I find it cuts ♏ 29. therefore I
ſay ♏ 29. is oppoſite to ♉ 29. Having found the oppoſite degree, I bring it into the *Weſt*, and alſo the Quadrant of Altitude, and joyn ♏ 29. to 18. degrees (accounted upwards on the
Quadrant) ſo ſhall ♉ 29. be depreſſed 18. degrees in the *Eaſt*
Side the Horizon : Then looking what Hour the Hour-Index points at in the Hour-Circle, I find it to be, 1. Hor. 8. Min.
which ſhews that Twilight begins at 8. Minutes paſt 1. a clock
in the Morning.

   And if you ſubſtract 1. Hour 8. Minutes, from 4. Hours 11.
Minutes, the time of Sun Riſing, found by the 7th. Probleme,
it leaves 3. Hours 3. Minutes for the length of Twilight : And
if you double 1. Hour 8. Minutes, the beginning of Twilight, it
makes 2. Hours 16. Minutes for the intermiſſion of Time between
Twilight in the Evening, and Twilight in the Morning. So that
May 10. abſolute Night is but 2. Hours 16. Minutes long, here
at *London*.

   The reaſon why you bring the degree oppoſite to the Suns
Place to the *Weſt*, is, becauſe the Quadrant containing but 90.
degrees will reach no lower then the Horizon; but this Probleme
requires it to reach 18. degrees beneath it : therefore by this
help, you have the Propoſition Anſwered, as well as if the Quadrant did actually reach 18. degrees below the Horizon. This
ſhift you may have occaſion to make in ſome other Problemes.

   If you would know when Twilight ends after Sun ſet; you
ſhall find it by bringing the degree of the *Ecliptick* oppoſite to
the Place of the Sun to 18. degrees of the Quadrant of Altitude,
on the *Eaſt* ſide the Horizon ; for then ſhall the Index of the
Hour-Circle point at 10. Hours 52. Minutes : which ſhews that
it continues Twilight till 52. Minutes paſt 10. a clock at Night,
May 10. here at *London*.

PROB. X.

The Suns Place given, to find its Amplitude; *And alſo
to know upon what point of the* Compaſs *it* Riſeth.

THe Globe &c. rectified : Bring the Suns Place to the *Eaſt*
Side the Horizon ; and the number of degrees intercepted
between

between the *East* point of the Horizon and the Suns Place, is the number of degrees of Amplitude that the Sun hath at its Rising; and bears its denomination either of *North* or *South*, according to its inclination to either point in the Horizon.

Or, if you would know upon what point of the Compass the Sun Rises, Look but in the Cirde of Winds; and against the Place of the Sun you have the name of the point of the Compass upon which the Sun Riseth.

Examples of both.

May 10. the Suns Place is ♉ 29. Therefore the Globe being rectified; I bring ♉ 29. to the *East* side the Horizon, and find it touch against 33, degrees 20. Minutes from the *East* point towards the *North* : Therefore I say the Sun hath *North* Amplitude 33, degrees 20. Minutes.

And to know upon what point of the Compass the Sun rises ; I keep the Globe to its Position, and look in the Cirde of Winds, in the outmost verge of the Horizon, and find the Suns Place against the Wind named *North East and by East* ; Therefore I say *May* 10. here at *London* the Sun riseth upon the *North East and by East* point of the Compass.

PROBL. XI.

The Hour of the Day given, to find the Heigth of the Sun.

THe Globe &c. Rectified. Turn about the Globe till the Index of the Hour-Cirde point (in the Hour-Circle) to the Hour of the Day : Then bring the Quadrant of Altitude to the Suns Place in the *Ecliptick* and the degree on the Quadrant which touches the Suns Place, shall be the number of degrees of the Suns Altitude.

Example.

May 10. here at *London* ; At 53. Minutes past 8. a clock in the Morning, I would know the Heigth of the Sun above the Horizon. Therefore I turn about the Globe till the Index of the

Hour-

Hour-Circle come to 53. Minutes paſt 8. a clock (which is almoſt
9.) in the Hour-Circle : And keeping the Globe to this Poſition, I bring the Quadrant of Altitude to the Suns place, *viz.*
♉ 29. (found by the third Probleme) and becauſe the Suns
Place touches upon 40. degrees of the Quadrant, therefore I ſay
May 10. 53. Minutes paſt 8. a clock in the Morning, here at
London, The Sun is juſt 40. degrees above the Horizon; or which
is all one, hath 40. degrees of Altitude.

PROB. XII.

The Altitude of the Sun, and Day of the Moneth given, to
find the Hour of the Day.

An hour defi-
ned.

AN Hour is the 24th. part of a Day and a Night, or the
ſpace of time that 15. degrees of the Equator takes up in
paſſing through the Meridian ; for the whole Equator
which contains 360. degrees, paſſes through the Meridian in 24. Hours, therefore 15. degrees which is the 24th. part
of 360, paſs through in one Hour. Theſe Hours are Vulgarly
divided into halfs, quarters, and half quarters; but Mathemati-
cally into Minutes, Seconds, Thirds, Fourths, &c. A Minute is

Minutes, Se-
conds, and
Thirds, &c.
defined.

the 60th. part of an Hour, ſo that 60, minutes make an Hour, 30,
half an Hour, 15. a quarter of an Hour : A Second is the 60th
part of a Minute : a third is the 60th part of a Second : a Fourth
is the 60th part of a Third: and ſo you may run on to Fifths,
Sixths, Sevenths, &c. if you pleaſe. 12. of theſe Hours make a
Day, and 12. more make a Night : ſo that Day and Night contain 24. hours as aforeſaid ; which are Vulgarly numbred
from Noon with 1, 2, 3, to 12, at Night ; and then begin
again with 1, 2, 3, till 12 at Noon : But by Aſtronomers they
are Numbred from Noon with 1, 2, 3, &c. to 12. at Night ;
and ſo forward to 13, 14, 15, till 24 ; which is juſt full Noon
the next Day. Yet in this Treatiſe I ſhall mention the Hours as
they are Vulgarly counted, *viz.* from 1. after noon, to 12. at
Night, and call the Hours after Midnight by 1, 2, 3, 4, &c. in
the Morning, to 12. at Noon again, the next Day. But to the
operation.

The Globe, &c. Rectified, Bring the Place of the Sun to the
Number of degrees of Altitude accounted upon the Quadrant of
Alt-

Altitude, and the Hour-Index fhall point at the Hour in the Hour-Circle : yet herein refpect muft be had to the Fore or After noons Elevation ; as fhall be fhewed in the next Probleme.

Example.

May 10. The Sun is elevated 40. degrees above the Horizon, here at *London* : Therefore having found the Place of the Sun, by the third Probleme, to be ♉ 29. I move the Globe and Quadrant till I can joyn the 29. degree of ♉ to the 40. deg, upon the Quadrant of Altitude ; and then looking on the Hour-Circle, I find the Index point at 53. Minutes paft 8. a clock, for the Fore noon Elevation ; and at 3. hours 7. Minutes for the After noons Elevation. Therefore if it be Fore-noon, I fay, It is 53. Minutes paft 8. a clock in the Morning. But if it be After noon I fay, It is 7. Minutes paft 3. a clock in the After noon.

PROB. XIII.

How to know whether it be Before or After Noon.

HAving made one Obfervation, you muft make a Second a little while after the Firft ; and if the Sun increafe in Altitude, it is Before Noon : but if it decreafe in Altitude, it is After Noon.

Example.

The Sun was at 8. hor. 53. Min. elevated 40. degr. above the Horizon : A little while after (fuppofe for examples fake a quarter of an hour,) *viz.* at 9. hor. 8. Min. I obferve again the heigth of the Sun, and find it 42. degrees high ; fo that the Altitude is increafed 2. degrees ; Therefore I fay, It is Fore-Noon : But if the Sun had decreafed in Altitude, I fhould have faid it had been After-Noon.

How to take Altitudes by the Quadrant, Aftrolabe, *and* Crofs-ftaff.

There are divers Inftruments whereby Altitudes may be taken : but the moft in ufe are the *Quadrant, Aftrolabe,* and *Crofs-ftaff.* A *Quadrant* is an Inftrument comprehended between two Straight lines making a Right Angle, and an

Arch

Arch difcribed upon the Right Angle, as on the Center, containing 90. degrees, which is a quarter of a Circle: and therefore
the Inftrument is called a *Quadrant.*   See this Figure.

A preprefents the Center ; upon which is faftned a Plumb-
line, A B the one fide, A C the other fide, upon which the Sights
are placed: B C the Arch or Quadrant, which is divided into 90.
equal parts, and numbred from B to C. D one Sight, E the
other Sight : F the Plumbet faftned to the Plumb-line.

When by this Inftrument you would obferve the heigth of
the Sun, you muft turn the Center A to the Sun, and let the
beams thereof dart in at the hole in the firft Sight D,
through the hole in the fecond Sight E; fo fhall the Plumb-line
ly upon the degree in the Limb, of the Suns Elevation : As if
the plumb-line ly upon the 20th degree, then fhall the Altitude be 20. degrees ; if on 25. the Altitude fhall be 25.
degrees : and fo for any number of Degrees the thred or Plumb-
line lies on, the fame number of Degrees is the Altitude of the
Sun.   But

But if it be a Star whoſe Altitude you would obſerve ; you muſt hold up the Quadrant, and joyn the Limb to your Cheek bone, and turn the Center towards the Star : then winking with one Ey, look through the holes of the Sights with the other Ey, till you can ſee the Star through thoſe holes ; ſo ſhall the Plumb-line (as before in the Sun) hang upon the degree in the Limb of the Stars Elevation.

Another ſort of Quadrants is made with a moveable Index, as is repreſented in this Figure.

A is the Center, A B and A C the two ſides, B C the Limb, D E two Sights fixed upon a moveable Index or Label ; F G two other Sights, for obſerving the Horizon.

When by this Quadrant you would obſerve an Altitude, the ſide B A muſt be parallel to the Horizon, and the Index muſt be moved till the Object (be it either the Sun Moon or any Star) be ſeen through the holes or ſlitts of the Sights placed on the Index; for then the Arch D B ſhall be the Elevation required. You

H   may

may know when the ſide B A is parallel to the Horizon, by obſerving the parting of Heaven from the Earth through the Sights
on the Side B A.

To take Altitudes by the Aſtrolabe.

The Aſtrolabe is a round Inſtrument, flat on either ſide,
upon one of the flats or Plains is diſcribed a Circle as B C D E,
divided into 360. equal parts or degrees, numbred from the
line of Level B A C, with 10, 20, 30, &c. to 90. in the Perpendicular D C. Upon the perpendicular is faſtned a Ring as F, ſo
as the Inſtrument hanging by it, the line of Level may hang parallel to the Horizon. Upon the Center is a moveable Label or
Ruler, as G H, whereupon is placed two Sights as I K.

If you deſire further inſtructions for making this Inſtrument,
you may peruſe *M*ʳ *Wright* in his *Diviſion of the whole Art*
of

of *Navigation,* annexed to his *Correction of Errors* : where he alſo ſhews the uſe of it at large ; which in brief is as follows.

You muſt hold the *Aſtrolabe* by the Ring in your left hand, and turning your right ſide to the Sun, lift up the Label with your right hand, till the beams of the Sun entring by the hole of the uppermoſt Vane or Sight, doth alſo pierce through the hole in the nethermoſt Vane or Sight ; and the deg. and part of deg. that the Label lies on is the height of the Sun above the Horizon.

But if it be a Star you would obſerve ; you muſt uſe the *Aſtrolabe* as you were directed to uſe the *Quadrant,* holding it up to your Cheek bone, and looking through the Sights, &c.

To take Altitudes by the *Croſs-ſtaff.*

This Inſtrument conſiſts of a Staf about a yard long, and three quarters of an inch ſquare : Upon it is fitted a Vane, (or ſometimes two, or three,) ſo as it may ſlide pretty ſtiff upon the Staff, and ſtand at any of the Diviſions it is ſet to.

The making is taught by Mr *Wright,* aforeſaid: But the uſe is as follows.

You muſt put that end of the *Croſs-ſtaff* which is next 90. degrees to your Cheek bone, upon the outter corner of your Ey, and holding it there ſteddy, you muſt move the Vane till you ſee the Horizon joyned with the lower end thereof, and the Sun or Star with the higher end ; then the degree and part of degree which the Vane cutteth upon the Staff, is the height of the Sun or Star.

Some of theſe waies for taking Altitudes have been formerly taught by others, that have treated upon the Uſe of Globes : and therefore becauſe ſome would be apt to think this Treatiſe un-compleat if I did not ſhew theſe waies alſo, I have thought fit to inſert them : Yet the ſame things may be performed by the Globe alone, without troubling your ſelf with multiplicity of In-ſtruments ; if your Globe be made with a hollow Axis ; for then if the Globe ſtand Horizontal, you ſhall by Obſerving the Object through the Axis have the degree of Elevation, noted by the ſuperficies of the Horizon.

PROB. XIV.

To obſerve with the Globe the Altitude of the Sun.

PLace the Globe ſo that the upper plain of the Horizon may ſtand parallel to the Plain of the Horizon of your Place; as was taught by the Second Probleme ; then turn the North Pole towards the Sun, and place it higher or lower, by moving the Meridian through the notches of the Horizon, till the beams of the Sun pierce quite through the Axis of the Globe : So ſhall the arch of the Meridian com-prehended between the Pole and the Horizon, be the number of Degrees that the Sun is elevated above the Ho-rizon.

Example.

March 20. juſt at noon, here at *London,* I would obſerve the Meridian Altitude of the Sun. Therefore placing the Ho-rizon Horizontal, as by the Second Probleme : I turn the North
Pole

Pole towards the Sun, and move it with the Meridian upwards or downwards, either to this fide or that, till I can fit it to fuch a Pofition that the Sun Beams may dart quite through the Axis of the Globe ; which when it does, I look on the Meridian and find 42. degrees 25. min. comprehended between the Pole and the fuperficies of the Horizon ; Therefore I fay the Meridian Altitude of the Sun *March* 20. here at *London,* is 42. degrees 25. min.

PROB. XV.

To find the Elevation of the Pole, by the Meridian Altitude of the Sun, and Day of the Moneth given.

THe Day of the Moneth is *March* 20. By the 4th Prob. you may find the place of the Sun to be ♈ 10. Therefore bring the Place of the Sun to the Meridian, and elevate it above the Horizon the fame number of degrees it hath in Heaven ; fo fhall the arch of the Meridian comprehended between the Pole and the Horizon, be the elevation of the Pole, in your Place.

Otherwife.

The Day of the Moneth given is *March* 20. fo that by the fourth Prob. you have the Suns Place ♈ 10; and by the fifth, the Declination of the Sun 3. 55. North : therefore the Declination being North, and you on the North fide the Equator ; you muft fubftract 3. 55. from the Meridian Altitude 42. 25. and there remains 38, 30. for the heighth of the Equinoctial above the Horizon ; but if your Declination had been South, you muft have added 3. 55. to the Meridian Altitude, and the Sum would have been the Elevation of the Equinoctial. Having the Elevation of the Equinoctial, you may eafily have the Elevation of the Pole ; for the one is alwaies the Complement of the other to 90. Thus the Height of the Equinoctial 38. 30. fubtracted from 90. leaves 51. 30. for the Elevation of the Pole, here at *London.* And thus it follows, that the Latitude of any Place from the Equinoctial, is alwaies equal to the Elevation of the

H 3 Pole :

Pole : for between the *Zenith* and the Equinoctial is contained the Complement of the Heighth of the Equinoctial above the Horizon to 90.

PROB. XVI.

To take the Altitude of any Star above the Horizon ; by the Globe.

T He Horizon of the Globe ſet parallel to the Horizon of the World, as before : Turn the North Pole towards the Star, and when you can ſee the Star through the Axis, the Northern notch of the Horizon will cut the degree of Elevation on the Meridian.

Example.

April 19. at 11. a clock at Night, I would obſerve the Altitude of *Spica Virgo* : Therefore I ſet the Horizon parallel to the Horizon of the World, as by the Second Probleme, and turn the Northern Pole till it point towards the Star : Then looking in at the South Pole of the Globe through the Axis, I ſhall ſee the Star, and have on the Meridian the Queſtion reſolved. But if it point not exactly, then I move the North Pole upwards or downwards, either to the right hand, or to the left, according as I may find occaſion, till I can ſee the Star through the Axis : and then the edge of the notch in the Horizon cuts 28. degrees 57. min. on the Brazen Meridian. Therefore I ſay *April* 19. at 11. a clock at Night, here at *London*, the Altitude of *Spica* ♍ is 30. degrees above the Horizon.

PROB. XVII.

By the Meridian Altitude of any Star given, to find the Height of the Pole.

J Oyn the Star to the Meridian, and place it to the Altitude obſerved ; ſo ſhall the number of degrees intercepted between the Pole and the Horizon, be the Elevation of the Pole.

Exam-

Example.

Spica Virgo is obferved to have 28. degrees 57. min Meridian Altitude; therefore I bring *Spica Virgo* to the Meridian, and raife it or deprefs it higher or lower as I find occafion, till it is juft 28. degrees 57. min. above the Horizon : Then I count the number of degrees between the Pole and the Horizon, and find them 51½. Therefore I fay the Elevation of the Pole is here at *London* 51½. Yet note, If the Star whofe Altitude you obferve have fewer number of degrees of Declination from the Pole, then the Elevation of the Pole, you may be apt to miftake in its coming to the Meridian ; for thofe Stars never fet; and therefore are twice Vifible in the Meridian in 24. hours, once above the Pole, and once under the Pole.

If your Star have greater Altitude then the North Star, it is above the Pole ; but if it have lefs, it is below the Pole : fo that if you know but whether it be above or below, it is enough ; for fo you may accordingly raife it to the Altitude on the Meridian it hath in Heaven, and joyn it to the Meridian either above or beneath the Pole, as the Star is placed in Heaven : and then the arch of the Meridian comprehended between the Pole and the Horizon, is the Elevation of the Pole, as aforefaid.

Otherwife.

Having the Meridian Altitude of the Star, you muft find its Declination by the 27. Probleme : and if the Declination be South, and you on the North fide the Equator, you muft ad the Declination to the Meridian Altitude, and the fum of both makes the Altitude of the Equinoctial : But if the Declination be North, and you on the North fide the Equator, you muft fubftract the Declination from the Meridian Altitude, (as was taught by the 15. Prob. in the Example of the Sun) and the remainder is the Altitude of the Equinoctial Then (as was taught by the 15 Probleme aforefaid) fubftract the Altitude of the Equinoctial from 90, the Remainder is the Elevation of the Pole in your Place.

Exam-

Example.

By the laſt Probleme the Meridian Altitude of *Spica Virgo* was 28 degrees 57 min, and the Declination of *Spica* by the 27th Probleme is found 9. degrees 33. min. South : therefore becauſe the Declination is South, I ad 9. degrees 33. min. to the Meridian Altitude, which makes 38. deg. 30. min. for the Elevation of the Equinoctial : which 38. deg. 30. min. ſubſtracted from 90. leaves 51. degrees 30. min. for the Elevation of the Pole here at *London.*

PROB. XVIII.

Another way to find the Height of the Pole by the Globe; if the Place of the Sun be given : and alſo to find the Hour of the Day, and Azimuth, and Almicantar of the Sun.

This muſt be performed by help of a *Spherick Gnomon,* (*as Blaew* calls it,) which is a ſmall Pin or Needle fixed perpendicularly into a ſmal Baſis with an hollow concave bottom, that it may ſtand upon the convexity of the Globe. Therefore the Horizon of the Globe being ſet parallel to the Horizon of the World, (as by the Second Probleme) the *Spherick Gnomon* muſt be ſet exactly upon the Place of the Sun; and then turning the Globe about (upon its Axis) either from *Eaſt* to *Weſt,* or contrarily from *Weſt,* to *Eaſt* ; or elſe by the Meridian, through the notches of the Horizon, till the *Spherick Gnomon* caſt no ſhadow on any ſide thereof; you have on the Meridian in the North point of the Horizon the number of degrees that the Pole is elevated above the Horizon.

Example.

Imagine the four Quarters of the Horizon of the Globe correſpond with the four Quarters of the Horizon of the World ; and the Plain of the Horizon of the Globe is parallel to the Plain of the Horizon of the World : The Suns Place is ♉ $29\frac{1}{4}$, which

which I find on the Globe, and place the *Spherick Gnomon* thereon ; Then at a guefs I move the Globe both on its Axis, and by the Meridian, (as neer as I can) fo as the *Spherick Gnomon* may caft no fhadow ; yet if it do, and the fhadow fall towards the North Pole; then I elevate the North Pole more, till the fhadow fals juft in the middle of it felf : but if the fhadow fall downwards, towards the South Pole, then I deprefs the North Pole : If the fhadow fall on the *Eaft* fide, I turn the Globe on its Axis more to the *Weft*; and if the fhadow fall to the *Weft*, I turn the Globe more into the *Eaft* : and the degree of the Meridian which the North point of the Horizon touches, is the degree of the Poles Elevation : which in this Example is $51\frac{1}{2}$. the Latitude of the City of *London*.

By this Operation you have alfo given the Hour of the Day in the Hour-Circle, if you keep the Globe unmoved : and the *Azimuth*, and *Almicantar*, if you apply but the Quadrant of Altitude to the Place of the Sun, as by the 22, and 23. Problemes.

PROB. XIX.

To obferve by the Globe the Diftance of two Stars.

YOu muft pitch upon two Stars in the Meridian ; and obferve the Altitude of one of them firft, and afterwards the Altitude of the other : Then fubftract the leffer Altitude from the greater, and the remainder fhall be the diftance required.

Example.

March 7. at 11. a clock at Night here at *London*, I fee in the Meridian the two Stars in the foremoft *Wheels* of the *Waggon*, in the Conftellation of the *Great Bear*, called by Sea-men the *Pointers*, (becaufe they alwaies point towards the Pole-Star.) Therefore to obferve the diftance between thefe two Stars, I firft obferve (as by the laft Probleme) the Altitude of the moft Northern to be 77. degree 59. minutes, and fet down that number of Degrees and minutes with a Pen and Ink on a Paper, or with a peece of Chalk or a Pencil on a Board : and afterwards I obferve the Altitude of the other Star which is un-

I   der

der it, as I did the firſt, to be 83. deg. 21. min. and ſet that
number of degrees and minutes alſo down, under the other
number of degrees and minutes : Then by ſubſtracting the leſſer
from the greater, I find the remainder to be 5. degrees 22. min.
which is the diſtance of the two Stars in the *Great Bear,* called
the *Pointers.*

PROB. XX.

*How you may learn to give a gueſs at the number of
degrees that any two Stars are diſtant from one
another ; or the number of degrees of Altitude the
Sun or any Star is elevated above the Horizon :
only by looking up to Heaven, without any Inſtrument.*

Between the *Zenith* and the Horizon is comprehended an
Arch of a Circle containing 90. degrees ; ſo that if you
ſee any Star in or neer the *Zenith,* you may know that
Star is 90. or neer 90. degrees high ; and by ſo much as
you may conceive it wants of the *Zenith,* ſo much you may
gueſs it wants of 90. degrees above the Horizon. By this Rule
you may gueſs at an Arch of Heaven containing 90. degrees, or
at an Arch of Heaven containing 45. degrees ; if by your imagination you divide the whole Arch into two equal parts, for
then ſhall each of them contain 45. degrees ; And if by your
imagination you divide the Arch of 90. into 3. equal parts, each
diviſion ſhall contain an Arch of 30. degrees, &c. But this way
is a little too rude for gueſſing at Stars elevated but few degrees,
or for Stars diſtant but few degrees from one another. Therefore
that you may learn to gueſs more preciſely at Diſtances in Heaven, you may either with a *Quadrant,* *Aſtrolabe,* or the *Globe* ;
find the exact diſtance of any two known Stars that are but few
degrees aſunder, and by a little revolving the diſtance of thoſe
Stars in your fancy, you may at length ſo imprint their diſtance
in your memory, that you may readily gueſs the diſtance of other
Stars by the diſtance of them.

Example.

You may find either by the *Globe, Quadrant,* or *Aſtrolabe,*
(for

(for they all agree) 3. degrees comprehended between the firſt Star in *Orions Girdle,* and the laſt ; therefore by a little rumi-nating upon that diſtance, you may imprint it in your fancy for 3. degrees, and ſo make it applicable to other Stars, either of the ſame diſtance, or more, or leſs : And the *Pointers* (by the laſt Probleme) are diſtant from one another 5. degrees and almoſt an half: Theſe are alwaies above our Horizon, and therefore may alwaies ſtand as a Scale for five and an half degrees ; So that by theſe for $5\frac{1}{2}$ degrees, and thoſe in *Orions Girdle* for 3. degrees, and others obſerved, either of greater or leſſer diſtance, you may according to your own Judgement ſhape a gueſs, if not exactly, yet pretty neer the matter of Truth, when you come to other Stars. Thus you may exerciſe your fancy upon Stars found to be 10. or 15. degrees aſunder, or more, or leſs ; and with a few experiments of this nature enure your Judgement to gueſs di-ſtances, and enable your memory to retain your Judgement.

This way of gueſſing will be exact enough for finding the Hour of the Night by the Stars, for moſt common Uſes ; or the Hour of the Day, by gueſſing at the Altitude of the Sun ; if after you have gueſſed at the Altitude, you ſhall work as was taught by Prob. 12. for the Hour of the Day : and as ſhall be taught in the next Probleme, for the Hour of the Night.

PROB. XXI.

*The Day of the Moneth, and Altitude of any Star given,
to find the Hour of the Night.*

THe Globe, Quadrant, and Hour Index rectified : Bring the Star on the Globe to the ſame number of Degrees on the Quadrant of Altitude that it hath in Heaven : So ſhall the Index of the Hour-Circle point in the Hour-Circle at the Hour of the Night.

Example.

March 10. the Altitude of *Arcturus* is 35. degrees above the Horizon, here at *London* : Therefore having the Globe,

Qua-

Quadrant and Hour Index rectified, I bring *Arcturus* on the Globe to 35. degrees on the Quadrant of Altitude : And then looking in the Hour-Circle, I find the Index point at 10. a clock ; which is the Hour of the Night.

PROB. XXII.

The Place of the Sun, and Hour of the Day given, to find its Azimuth in any Latitude affigned.

THe Globe, &c. rectified to your Latitude : Turn the Globe till the Index of the Hour-Circle come to the given hour ; and bring the Quadrant of Altitude to the Place of the Sun : fo fhall the number of degrees contained between the Eaft point of the Horizon and the degree cut by the Quadrant of Altitude on the Horizon, be the number of degrees of the Suns Azimuth, at that time.

Example.

May 10. at 53. minutes paft 8. a clock in the Morning, I would know the Azimuth of the Sun : Therefore (the Globe being firft rectified) I turn about the Globe till the Index of the Hour-Circle point to 53. minutes paft 8. a clock, or which is all one, within half a quarter of an hour of 9 ; then I move the Quadrant of Altitude to the degree the Sun is in that Day, and there let it remain till I fee how many degrees is contained between the *North* point and the Quadrant ; which in this Example is 108. deg. 25. min. And becaufe this diftance from the *North*, exceeds 90. degrees ; therefore I fubftract 90. degrees from the whole, and the remains is 18. degrees 25. min. for the Azimuthal diftance of the Sun from the *Eaft* point towards the *South*. But if it had wanted of 90. degrees from the North point, then fhould the Complement of 90. have been the Azimuthal diftance of the Sun from the Eaft point.

PROB. XXIII.

The Place of the Sun, and hour of the Day given, to find the Almicantar of the Sun.

THe Almicantars of the Sun is upon the matter the fame thing with the Altitude of the Sun : only with this diftinction. The Almicantars are Circles parallel to the Horizon, difcribed by the degree of the Quadrant of Altitude upon the *Zenith* as its Center, by turning the Quadrant round about the Globe till it comes again to its firft Place : But the Altitude is an Arch of the Vertical Circle, comprehended between the Horizon and any point of the Globe affigned. Their agreement confifts in this ; When the Sun or any Star has any known Almicantar, they are faid to have the fame number of degrees of Altitude ; As if the Sun be in the 20th Almicantar ; he hath 20 degrees of Altitude ; if in the 30th Almicantar, he hath 30. degrees of Altitude, &c. Now becaufe the Operation is the fame for finding the Altitude and Almicantar, I fhall refer you to the 11th Probleme ; which fhews you how to find the Altitude or Heighth; and by confequence the Almicantar.

PROB. XXIV.

The Place of the Sun given, to find what Hour it comes to the Eaft, *or* Weft, *and what Almicantar it then fhall have.*

THe Globe, Quadrant, and Hour Index rectified, Bring the Quadrant of Altitude to the *Eaft* point in the Horizon, if you would know what hour it comes to the *Eaft;* or to the *Weft* point, if you would know what hour it comes to the *Weft* : Then turn about the Globe till the place of the Sun come to the Quadrant of Altitude ; and the Index of the Hour Circle fhall point at the hour of the Day : which on the Day aforefaid will be 7. hor. 7 min. in the Morning, that the Sun commeth to the *Eaft,* and 4 hor. 53. min. after noon, that the Sun commeth to the Weft. And if you then count the number of degrees from

I 3   the

the Horizon upwards on the Quadrant of Altitude, it will ſhew you the Almicantar of the Sun for that time ; which will both Morning and Evening be 15, deg. 30. min. as was taught you by the laſt Probleme.

PROB. XXV.

To know at any time what a clock it is in any other Part of the Earth.

THe difference of Time is reckoned by the acceſs and pro-greſs of the Sun : for the Sun gradually circumvolving the Earth in 24. hours, doth by reaſon of the Earths rotundity en-lighten but half of the ſphear at one and the ſame moment of Time ; as ſhall be ſhewed hereafter : ſo that hereby it comes to paſs, that when with us here in *England* it is 6. a clock in the Morning, with thoſe that have 90. degrees of Longitude to the Weſtward of us, it is yet Midnight : with thoſe that have 180. degrees of Longitude from us, it is Evening ; And with thoſe that have 90. degrees of Longitude to the *Eaſtwards*, it is Noon. So that thoſe to the *Eaſtward* have their Day begin ſooner then ours : But to the *Weſtward* their Day begins after ours. Therefore that you may know what Hour it is in any Place of the Earth, of what diſtance ſoever it be, you muſt firſt Bring the Place of your own Habitation to the Meridian, and the Index of the Hour Circle to 12. on the Hour Circle; Then bring the other Place to the Meridian, and the Arch of the Hour Circle comprehended between the hour 12. and the In-dex, is the difference in Time between the two Places.

Example.

London in *England*, and *Surat* in the *Eaſt Indies* : Firſt I bring *London* to the Meridian, and turn the Index of the Hour-Circle to 12 ; then I turn the Globe *Weſtward*, becauſe *London* is *Weſtward* of *Surat*, till *Surat* come to the Meridian ; and ſee at what Hour the Index of the Hour Circle points, which in this Example is 5. hor. 54. minutes : And becauſe *Surat* lies to the *Eaſtward* of us ſo many degrees, therefore as was ſaid before,

their

their Day begins fo much before ours : So that when here at *London* it is 6. a clock in the Morning, at *Surat* it will be 11. a clock 54. minutes ; when with us it is 12. a clock, with them it will be 5 a clock 54. minutes afternoon.

If you would know the difference of Time between *London* and *Jamaica* ; Working as before, you may find 5. hor. 15. min. But *Jamaica* is to the *Weft* of *London* ; therefore their Day begins 5. hor. 15. min. after ours : fo that when with us it is Noon, with them it will be but three quarters of an hour paft 6. a clock in the Morning : and when with them it is Noon, with us it will be one quarter paft 5. a clock after Noon, &c.

Or you may yet otherwife know the difference of Time, if you divide the number of Degrees of the Equinoctial that pafs through the Meridian while the Globe is moved from the firft Place to the fecond, by 15. fo fhall the product give you the difference of hours and minutes between the two Places : as you will find if you try either of thefe Examples, or any other.

PROB. XXVI.

To find the Right Afcenfion of the Sun, or Stars.

THe Right Afcenfion of any point on the Globe is found by bringing the point propofed to the Meridian, and counting the number of degrees comprehended between the *Vernal Colure,* and the Meridian.

Example, for the Sun.

June 1. I would know the Right Afcenfion of the Sun : His Place found, as by the third Probleme, is ♊ 20. Therefore I bring ♊ 20. to the Meridian ; and then the Meridian cuts the Equinoctial in 79. degrees 15. minutes, accounted from the *Vernal* point ♈ : Therefore I fay the Right Afcenfion of the Sun *June* 1. is 79. deg. 15. Minutes.

Example, for a Star.

I take *Capella,* alias *Hircus,* the *Goat* on *Auriga's fholder,*

and

and bring it to the Meridian ; and find the Meridian cut the
Equinoctial (counting as before from the *Vernel* point ♈) in 73.
degrees 58. minutes : Therefore I ſay, the Right Aſcenſion of
Hircus is 73. degrees 58. min. Do the like for any other point of
the Globe propoſed.

PROB. XXVII.

To find the Declination of the Sun, or Stars.

THe Declination of any point on the Globe is found by
bringing the point propoſed to the Meridian, and counting
the number of degrees comprehended on the Meridian between
the Equinoctial and the point propoſed : and bears its Denomination of *North* or *South,* according as it is ſcituate on the *North*
or *South* ſide the Equinoctial.

Example, for the Sun.

June 1. I would know the Declination of the Sun. His Place
found, as before, is ♊ 20. Therefore I bring ♊ 20. to the Meridian ; and find 23. degrees 8. min. comprehended on the Meridian between the Equinoctial and ♊ 20. and becauſe ♊ is on
the *North* ſide the Equinoctial ; Therefore I ſay, *June* 1. The
Sun hath *North* Declination 23. degrees 8. minutes.

Example, for a Star.

I take *Hircus* aforeſaid, and bring it to the Meridian, and
find 45. degrees 40. minutes comprehended on the Meridian
between the Equinoctial and the Star *Hircus.* And becauſe
Hircus is on the *North* ſide the Equinoctial; Therefore I ſay,
Hircus hath *North* Declination 45. degrees 40. min. Do the
like for any other point on the Globe propoſed.

But Note, The Right Aſcenſion and Declination of the Sun alters dayly; for in twelve Moneths he runs through every degree
of Right Aſcenſion, and in three Moneths to his greateſt Declination: But the Right Aſcenſion and Declination of the Stars is
ſcarce perceiveable for ſome Years : Yet have they alſo an alteration of Right Aſcenſion and Declination: For, thoſe Stars
that

that have but few degrees of Right Afcenfion, will in procefs of Time have many ; and thofe Stars between the Tropick that have North Declination, will in length of Time have South Declination ; and the contrary (as fhall be more fully fhewed hereafter:) For, the Stars moving upon the Poles of the *Ecliptick* go forwards in Longitude one whole Degree in 70 ½ Years (as hath been fhewed before, *Book* 1. *Chap.* 3. *Sect.* 3.) and fo alter both their Right Afcenfion, and Declination ; as may be feen by this following Table of Right Afcenfions and Declinations of 100. of the moft eminent fixed Stars, Calculated by *Tycho Brahe*, for the Years 1600. and 1670. which I have inferted ; partly, becaufe by it you may fee the differences of their Right Afcenfions and Declinations in 70 ½ Years ; and partly to Accomodate thofe that may have occafion to know their Right Afcenfions and Declinations neerer than the Globe can fhew them.

A Table of the Right Afcenfions and Declinations of 100. Select fixed Stars ; Calculated by Tycho Brahe, *for the Years 1600, and 1670. As alfo their Difference of Right Afcenfions and Declinations, in 70. Years.*

Names of the Stars.	1600				Differentia.				1700			
	R. Afc.		Declin.		R. Af.		Decl.		R. Afc.		Declin.	
Scedir, *in Caffiopeæ.*	4	36	54	21	N	1	22	34	S	5	58	54	55
The Pole Star. (tail.	5	47	87	9½	N	3	59	34	S	9	46	87	43½
Southern *in the whales*	5	51	20	12	S	1	17	34	N	7	8	19	38
Caffiopeæ's Belly.	8	21	58	33	N	1	27	34	S	9	48	59	7
Girdle *Andromeda.*	11	50	33	32	N	1	23	33	S	13	13	34	5
Knee of *Caffiopeæ.*	15	3	58	7	N	1	35	33	S	16	38	58	40
1. *in* ♈ horn.	22	56	17	19	N	1	23	31	S	24	19	17	50
Whales belly.	22	59	12	16	S	1	15	31	N	24	14	11	45
2. *in* ♈ horn.	23	10	18	50	N	1	22	31	S	24	32	19	31
South foot of *Andromeda.*	24	55	40	23	N	1	29	30	S	26	24	40	53

K

Names of the Stars.	1600			Differentia.		1700	
	R.Asc.	Declin.		R.As.	Decl.	R.Asc.	Declin.
In the Knot in the line ♓.	25 22	0 50 N		1 18	30 S	26 40	1 20
* Star in ♈ head.	26 23	21 33 N		1 25	30 S	27 38	22 3
* In the whales jaw.	40 25	2 29 N		1 15	25 S	41 40	2 54
Caput Medusæ	40 38	39 22 N		1 37	25 S	42 15	39 47
* In Perseus side.	44 2	48 22 N		1 28	21 S	45 30	48 43
* In the Pleiades.	50 57	22 49 N		1 29	21 S	52 26	23 10
In the Nostrils of ♉.	59 16	14 37 N		1 25	17 S	60 41	14 54
North Ey of ♉.	61 21	18 14 N		1 24	17 S	62 45	18 31
Aldebaran.	63 16½	15 38 N		1 26½	15 S	64 43	15 53
Hircus, Capella.	71 49	45 30 N		1 49	10 S	73 38	45 40
* Orions foot, Rigel.	73 51½	8 43 S		1 15½	9½ S	75 7	8 33½
North Horn ♉.	75 16	28 12 N		1 37	8 S	76 53	28 20
Orions left sholder.	75 58	5 55 N		1 19	8 S	77 17	6 3
Belly of the Hare	77 48	21 6 S		1 5	7 N	78 53	20 59
1. In Orions Girdle	77 58	0 39 S		1 17	7 N	79 15	0 32
Uppermost in Orions face.	78 21	9 36 N		1 22	7 S	79 41	0 43
South Horn ♉.	78 26	20 51 N		1 31	7 S	79 57	20 58
2. In Orions Girdle.	79 1	1 30 S		1 17	6 N	80 18	1 24
Last in Orions Girdle.	80 10	2 12 S		1 16	5 N	81 26	2 7
Auriga's right Sholder.	82 40	44 50 N		1 55	4 S	84 35	44 54
Orions right Sholder.	83 26	7 16 N		1 22	4 S	84 48	7 20
* Foot ♊.	93 38	16 40 N		1 28	2 N	95 6	16 38
Great Dog Sirius. (Twin.	96 53	16 11 S		1 7	4 N	98 0	16 15
Head of Castor, the first	107 9	32 41 N		1 44	11 N	108 53	32 30
The little Dog, Procyon.	109 37	6 12 N		1 20	12 N	110 57	6 0
Head Pollux, second Twin.	110 13	28 55 N		1 34	12 N	111 47	28 43
* In the Stern of the	117 39	23 11 S		1 4	15 S	118 43	23 26
Præsepe ♋ (Ship.	124 20	21 2 N		1 28	19 N	125 48	20 43
Northern Asse ♋	124 58	22 51 N		1 30	20 N	126 28	22 31
Southern Asse ♋	125 27	19 35 N		1 27	20 N	126 54	19 15
The Heart of Hydra.	137 1	6 57 S		1 15	25 S	138 16	7 22
South of 3. in neck ♌	146 22	18 42 N		1 28	28 N	147 50	18 14
Lions Heart, Basiliscus.	146 45½	13 53½ N		1 53½	28½ N	148 8	13 25
North of 3. in neck ♌	148 33	25 23 N		1 23	29 N	150 1	24 54
Middle of 3. in neck ♌	140 25½	21 50 N		1 50	29 N	150 51	21 21

Names of the Stars.	1600 R.Asc.		Declin.			Diff. R.Aſ.		Decl.		1700 R.Asc.		Declin.	
Firſt loweſt in □ *Urſa Maj.*	159	12	58	31	N	1	37	32	N	160	49	57	59
Firſt upper in □ Dubbe	159	37	63	54	N	1	41	32	N	161	18	63	22
✳ back ♌.	163	10	22	43	N	1	27	34	N	164	37	22	9
Lions tail. (*Major.*	172	9	16	49	N	1	19	34	N	173	28	16	15
following loweſt in □ *Urſa*	173	3	55	57	N	1	23	34	N	174	26	55	23
Uppermoſt following in □.	178	50	59	15	N	1	20	34	N	180	10	58	41
Girdle ♍.	188	53	5	37	N	1	18	34	N	190	11	5	3
Rump *Urſa Major*, Aliot.	189	1	58	10	N	1	19	33	N	190	10	57	37
Vindemiatrix, ♍.	190	36	13	8	N	1	17	33	N	191	53	12	35
Spica ♍.	196	4	9	1	S	1	19½	32½	S	197	23½	9	33½
Middle tail Urſa Major.	196	54	57	3	N	1	3	32	N	197	57	56	31
End tail *Urſ. Major.*	202	54	51	22	N	1	2	31	N	203	56	50	51
Arcturus.	209	23½	21	18½	N	1	11	29½	N	210	34½	20	49
Left Sholder of *Bootes.*	214	2	40	3	N	1	2	27	N	215	4	39	36
South Scale ♎.	217	14½	14	18	S	1	23	37	S	218	37½	14	45
North Scale ♎.	223	54½	7	50	S	1	21½	24	S	225	16	8	14
✳ Northern Crown.	229	26	28	6	N	1	5	21	N	230	31	27	45
✳ Serpents neck.	231	12	7	46	N	1	15	21	N	232	27	7	25
Northern of 3. ✳ in front	235	34	18	38	S	1	28	19	S	237	2	18	57
Left hand *Ophiucus.* (♏	238	25	2	37	S	1	23	18	S	239	48	2	55
Heart ♏. Antares,	241	18	25	26	S	1	32	16	S	242	50	25	42
Right Shold. *Hercules.*	243	15	22	27	N	1	5	15	N	244	20	22	12
Left knee of *Ophiucus.*	243	49	9	39	S	1	23	15	S	245	12	9	54
Right knee of *Ophiucus.*	251	50	15	7	S	0	50	10	S	252	40	15	17
Head of *Hercules.*	254	6	14	55	N	1	8	8	N	255	14	14	47
Left Sholder of *Hercules.*	254	40	25	22	N	0	52	8	N	255	32	25	14
Head of *Ophiucus.*	259	5	12	56	N	1	11	7	N	260	16	12	49
Right Sholder of *Ophiucus.*	260	56	4	49	N	1	13	5	N	262	9	4	44
✳ head of the Dragon.	266	52	51	37	N	0	35	2	N	267	27	51	35
✳ Lyræ.	275	52	38	28	N	0	50	4	S	276	42	38	32
Moſt Eaſtern in Head ♐.	281	32	21	35	S	1	31	8	N	283	3	21	27
Vultures tail.	281	47	13	20	N	1	13	8	S	283	0	13	28
In the Swans Beak.	288	40	27	10	N	1	1	11	S	289	41	27	21
✳ in Vulture.	292	49	7	54	N	1	17	13	S	294	6	8	7
In the Swans North wing.	293	10	44	12	N	0	48	14	N	293	58	44	26

Names of the Stars.	1600			Differentia.			1700	
	R.Aſc.	Declin.		R.Aſ.	Decl.		R.Aſc.	Declin.
Upper horn ♑.	289 57	13 40	S	1 25	16	N	300 22	13 24
Lower horn ♑.	299 39	15 57	S	1 27	17	N	301 6	15 40
In the Swans breaſt.	302 1½	39 1	N	0 53½	18	S	302 55	39 19
Left hand of ♒.	306 32	10 53	S	1 16	19	N	307 48	10 34
Swans Tail.	306 57½	43 53½	N	0 51½	20½	S	307 49	44 14
In the Swans South wing.	307 31	32 30	N	1 0	21	S	308 31	32 51
Left Sholder ♒.	317 37	7 15	S	1 21	26	N	318 58	6 49
1. In tail ♑.	319 28	18 21	S	1 26	26	N	320 54	17 55
In Cepheus Girdle.	320 46	68 50	N	0 22	26	S	321 8	69 16
In Pegaſus mouth.	321 10	8 5	N	1 18	26	S	322 28	8 31
2. In tail ♑.	321 16	17 51	S	1 25	27	N	322 41	17 24
Right Sholder of ♒.	326 19	2 13	S	1 20	29	N	327 39	1 44
Fomahant, ♒.	338 46	31 39	S	1 25	31	N	340 11	31 8
Scheat. Pegaſus.	241 9	25 56	N	1 12	32	S	342 11	26 28
Marchab. Pegaſus.	341 15	13 5	N	1 15	32	S	342 30	13 37
Mouth of Southern fiſh.	344 9	1 7	N	1 17	33	S	345 26	1 40
Head of Andromeda.	356 59	26 54	N	1 17	34	S	358 16	27 28
* Caſſiopeæ's chair.	357 5	56 58	N	1 15	34	S	358 20	57 32
End of Pegaſus wing. (tail	358 14	12 58	N	1 16	34	S	359 30	13 32
Northern in the whales	359 49	11 1	S	1 18	34	S	1 7	10 27

The Vſe of this Table.

The firſt Collumne on the left hand is the names of the Stars. The Second Collumne ſhews the degrees and minutes of *Right Aſcenſion*, for the Year 1600. The third the *Declination* for the ſame Year. The fourth ſhews whether the Declination be *North* or *South* ; N ſtands for *North*, S for *South*. The fifth ſhews the difference in degrees and minutes of *Right Aſcenſion* of the Stars, between the Years 1600. and 1670. The ſixth ſhews the Difference of *Declination* ; and whether it be *North*, or *South*. The ſeventh ſhews the *Right Aſcenſion* in degrees and minutes, for the Year 1670. The eighth ſhews the *Declination* in degrees and minutes for the ſame Year.

By this Table you may perceive the fixed Stars increaſe in Right Aſcenſion, till they come to the *Vernal* Colure ; from
whence

whence the number of their Right Afcenfion is reckoned : and
by the Collumne of their Difference of Right Afcenfion, you
may fee how much they increafe in 70. Years. And if you
would know how much they increafe for any other number of
Years, you muft find what proportion they have to 70, and the
fame proportion the Difference of the Right Afcenfion of the
Stars will have to the Difference in the Table.

Example.

I would know the Difference of Right Afcenfion the *Pole-
Star* will have in 35. Years. I find in the fifth Collumne the
Difference of Right Afcenfion of the *Pole Star* to be 3. degrees
59. min. Therefore by the Rule of Proportion, I fay, If 70.
Years give 3. degrees 59. min. 35. Years fhall give 1. degree
59½. min: and fo proportionably for any other number of
Years.

Though this Rule ferves for finding the Difference of Right
Afcenfion of any Star ; Yet it will not ferve for finding the
Difference of any Stars Declination. For the Stars on the
North fide the Equinoctial between the *Hyemnal* and *Solsticial*
Colures, and on the South fide the Equinoctial between the
Solsticial and *Hyemnal* Colures, increafe in Declination. But
the Stars on the South fide the Equinoctial between the *Hyemnal* and *Solsticial* Colures, and on the North fide the Equino-
ctial between the *Solsticial* and *Hyemnal* Colures, Decreafe in
Declination : as you may yet more plainly fee by the Globe, if
you bring 66½ deg. of the Meridian to the North fide of the Horizon, and fcrew the Quadrant of Altitude to 66½ degrees in the
Zenith, and Declination of the Pole of the *Ecliptick* ; and bring
the *Hyemnal* Colure to the Meridian ; for fo fhall the Pole of
the *Ecliptick* be joyned with the center of the Quadrant of
Altitude, and the *Ecliptick* with the Horizon ; and all the
Circles that the feveral degrees on the Quadrant make in a Revolution from *Weft* to *Eaft* upon the Poles of the Ecliptick, reprefent the great Revolution of every Star that each degree on
the Quadrant cuts. And thus demonftratively will be reprefented the progrefs of the fixed Stars through every degree of Longitude, and by confequence the alteration of their Right Afcenfion, and Declination. For, Imagining that degree of the Qua-
K 3   drant

drant of Altitude to be the Star, which juſt reaches the Star ;
you may by turning about the Quadrant, ſee how Obliquely
the Star (or the degree repreſenting the Star) either moves
about, or cuts the Equinectial, and all Circles parallel to the
Equinoctial ; and thereby obſerve it ſometimes to incline in motion to, and other times to decline in motion from the Equinoctial. But how long time it will be ere the Star inclines to,
or declines from the Equinoctial, you may know by finding
the diſtance of Longitude in degrees it hath from either the
Solſticial or *Hyemnal* Colure; and with reſpecting the foregoing Rules in its Poſition, you may by the Table in Book 1,
Chap. 3. Sect. 3. ſatiſie your ſelf.

Example.

The moſt Northerly Star in the *Girdle of Orion* doth yet decreaſe in Declination. But I would know how long it ſhall decreaſe ; Therefore by the 32. Probleme, I find the Longitude
of that Star to be for the Year 1670. 77. deg. 51. min. which
ſubducted out of 90, (the diſtance of the Solſticial Colure from
the Equinoctial,) leaves 12. 9, for the diſtance of that Star from
the Solſticial Colure. Therefore by the Table aforeſaid, I find
what number of Years anſwers to the motion of 12. deg. 9.
min. And becauſe I cannot find exactly the ſame number of
degrees and minutes in the Table, I take the number neereſt to
it ; which is 14. degrees 10. minutes, and that is the motion of
the Ecliptick in 1000. Years. But becauſe this 14. degrees 10.
minutes is 2. degrees 1. minute too much, I ſeek 2. degrees,
1. min. in the Table, and the number of Years againſt it I would
ſubduct from the number of Years againſt 14. deg. 10. min.
and the remainder would be the number of Years required: But
2. deg. 1. min. I cannot find neither, therefore I muſt take the
number of degrees and minutes neereſt to it, which is 2. deg. 50.
min. and that yeelds 200. Years ; which ſubducted out of
1000. leaves 800. Years. But becauſe this is alſo too much by
the motion of 49. min. Therefore I ſeek for 49. min. in the
Table, and ſubduct the number of Years againſt it from 800,
and the remainder would be the number of Years required. But
49. min. is not in the Table neither, Therefore I take the neereſt
to it, which is 51. min. and that yeelds 60. Years; which ſubducted out of 800. leaves 740. But this is likewiſe too much by
the

the motion of two min. Therefore I feek 2. min. in the Table, but cannot find it neerer then 2½, and againft it I find 3. Years, which 3. Years I fubduct out of 740, and the Remainder is 737. the number in Years required. You may if you pleafe for exactnefs, fubduct for the ½ min. 8. Moneths ; fo have you 736. Years 4. Moneths, for the Time that the moft Northerly Star in the *Girdle of Orion* will decreafe in Declination after the Year 1670. which will be till *An. Dom.* 406. after which time it will increafe in Declination for 12706. Years together, till it come to have 47. degrees 8. min. of Declination : at which time it will be in or very neer the place of the moft Southerly Star of the *Southern Crown* ; and that Star in its place.

And thus the Pole Star is now found to increafe in Declination, and will yet this 421 Years : after which time it will decreafe in Declination for 12706 Years together, till it come to be within 42. degrees 42. minutes of the Equinoctial, in the void fpace now between *Draco* and *Lyra* ; at which time *Lyra* will be almoft as neer the Pole, as the Pole Star now is ; and then the moft proper to be the Northern Pole Star : And the laft Star in the *Stalk of the Doves mouth* will be then very neer the Southern Pole, and therefore moft fit to be the Southern Pole-Star.

PROB. XXVIII.

The Place of the Sun or any Star given, to find the Right Defcenfion, and the Oblique Afcenfion, and the Oblique Defcenfion.

BRing the Place of the Sun or the Star to the Meridian under the Horizon, and the degree of the Equator that comes to the Meridian with it is the Degree of Right Defcenfion.

For the Oblique Afcenfion.

Bring the Place of the Sun or the Star to the Eaft fide the Horizon, and the degree of the Equator cut by the Horizon, is the Degree of Oblique Afcenfion of the Sun or Star.

For

For the Oblique Defcenfion.

Bring the Place of the Sun or Star to the Weft fide the Horizon, and the degree of the Equinoctial cut by the Horizon is the Degree of Oblique Defcenfion. They need no Examples.

PROB. XXIX.

Any Place on the Terreftrial Globe being given, to find its Antipodes.

B Ring the given Place to the Meridian, fo may you (as by the firft Probleme) fee its Longitude and Latitude ; then turn about the Globe till 180. degrees of the Equator pafs through the Meridian; and keeping the Globe to this Pofition, number on the Meridian 180. degrees from the Latitude of the given Place : and the point juft under that degree is the *Antipodes.*

Example.

I would find the *Antipodes of Cuida Real,* an Inland Town of the *Weft Indies,* which lies upon the River *Parana,* an Arm of *Rio de la Plata* : Therefore I bring *Cuida Real* to the Meridian, and find (as by the firft Probleme) its Latitude 23. 50: South; and its Longitude 333. degrees : Then I turn about the Globe till 180. degrees of the Equator pafs through the Meridian ; and keeping the Globe to that pofition, I number fo many degrees North Latitude as *Parana* hath South, *viz.* 23. 50. and juft under that degree I find *Lamoo,* a Town lying upon the Coaft of *China,* in the Province of *Quancij* : Therefore I fay *Lamoo* is juft the *Antipodes of Cuida Real.*

Another way.

Bring the given Place to the North or South point of the Horizon, and the point of the Globe denoted by the oppofite point of the Horizon, is the *Antipodes* of the given Place.

PROB.

PROB. XXX.

To find the Perecij *of any given* Place, *by the* Terreſtrial *Globe.*

BRing your Place to that ſide the Meridian which is in the South notch of the Horizon, and follow the Parallel of that Place on the Globe till you come to that ſide the Meridian which is in the Northern notch of the Horizon ; and that is the *Perecij* of your Place.

PROB. XXXI.

To find the Antecij *of any given* Place, *upon the* Terreſtrial *Globe.*

B Ring your Place to the Meridian, and find its Latitude by the firſt Probleme ; If it have North Latitude, count the ſame number of degrees on the Meridian from the Equator Southwards ; But if it have South Latitude, count the ſame number of degrees from the Equator Northwards : and the point of the Globe directly under that number of degrees is the *Antecij* of your Place.

PROB. XXXII.

To find the Longitude *and* Latitude *of the Stars, by the* Coeleſtial *Globe.*

THe Quadrant of Altitude will reach but 90. degrees, as was ſaid Prob. 9. Therefore if the Star you enquire after be on the North ſide the Ecliptick, you muſt elevate the North Pole 66 ½ degrees above the North ſide the Horizon : If on the South ſide the Ecliptick, you muſt elevate the South Nole 66 ½ degrees above the South ſide the Horizon : Then bring the *Solſticial Colure* to the Meridian on the North ſide the Horizon, and ſcrew the Quadrant of altitude to the *Zenith,* which will be in 23 ½ degrees from the Pole of the World : So ſhall the Ecliptick ly in the Horizon, and the Pole of the Ecliptick alſo ly under the Center of the Quadrant of Altitude (as was

L ſhewed

ſhewed Prob. 27.) Now to find the Longitude of any Star, do thus Turn the Quadrant of Altitude about till the graduated edge of it ly on the Star ; and the degree in the *Ecliptick* that the Quadrant touches is the Longitude of that Star.

Example, for a Star on the North ſide the *Ecliptick.*

I would know the Longitude of *Marchab,* a bright Star in the wing of *Pegaſus :* I find it on the North ſide the *Ecliptick,* Therefore I elevate the North Pole, and placing ♋ on the North ſide the Meridian, I ſcrew the Quadrant of Altitude to the *Zenith,* as aforeſaid : Then laying the edge of the Quadrant of Altitude upon that Star, I find that the end of it reaches in the *Ecliptick* to ♓ 18. 56. Therefore I ſay, the Longitude of *Marchab* is ♓. 18. 56.

For the Latitude of a Star.

The Degree of the Quadrant of Altitude that touches the Star is the Latitude of the Star.

Example.

The Globe and Quadrant poſited as before, I find 19. deg. 26. min. (accounted upwards on the Quadrant) to touch *Marchab* aforeſaid : Therefore I ſay, the Latitude of *Marchab* is 19. deg. 26. min.

And thus by elevating the South Pole and placing the Globe and Quadrant of Altitude as aforeſaid, I ſhall find *Canicula* have 15. degrees 57. min. South Latitude, and 21. degr. 18. min in ♋, Longitude.

PROB. XXXIII.

To find the Diſtance between any two Places, on the Ter-reſtrial Globe.

THis may be performed either with the Quadrant of Al-titude, or with a pair of Compaſſes : with the Qua-drant of Altitude, thus: Lay the lower end thereof to one Place, and ſee what degree reaches the other Place, for
that

that is the number of degrees between the two Places. If you multiply that number of Degrees by 60 the Product ſhall be the number of Engliſh Miles between the two Places.

Example.

I would know the diſtance between *London* and the moſt Eaſterly point of *Jamaica*; I lay the lower end of the Quadrant of Altitude to *Jamaica,* and extending the other end towards *London,* I find 68½. deg. comprehended between them: Therefore I ſay 68½ is the number of degrees comprehended between *London* and *Jamaica.*

If you would find the Diſtance between them with your Compaſſes, you muſt pitch one foot of your Compaſſes in the Eaſt point of *Jamaica,* and open your Compaſſes till the other foot reach *London* ; and keeping your Compaſſes at that Diſtance apply the feet to the Equinoctial line, and you wil find 68½ degrees comprehended between them : as before.

If you multiply 68½. by 60, is it gives 4110. *Engliſh* miles.
If you multiply it by 20, it gives 1370. *Engliſh* Leagues.
If you multiply it by 17½, it gives 1199. *Spaniſh* Leagues.
If you multiply it by 15, it gives 1054 *Dutch* Leagues.

PROB. XXXIV.

To find by the Terreſtrial Globe upon what point of the Compaſs any two Places are ſcituate one from another.

FInd the two Places on the Terreſtrial Globe, and ſee what Rumb paſſes through them; for that is the point of the Compaſs they bear upon.

Example.

Briſtol and *Bermudas* are the Places: I examine what Rhumb paſſes through them both: and becauſe I find no Rhumb to paſs immediately through them both, Therefore I take that Rhumb which runs moſt Parallel to both the Places ; which in this Example is the tenth Rhumb counted from the North towards the left hand ; and is called as you may ſee by this following

Figure

Figure *Weſt South Weſt* ; Therefore I ſay *Bermudas* lies ſcituate from *Briſtol Weſt South Weſt* ; and by contraries *Briſtol* lies ſcituate from *Bermudas Eaſt North Eaſt.*

PROB. XXXV.

To find by the Coeleſtial Globe the Coſmical Riſing *and* Setting *of the Stars.*

WHen any Star Riſes with the Sun, it is ſaid to Riſe *Coſmically.*

And when any Star Sets when the Sun Riſes, it is ſaid to Set *Coſmically.*

To find theſe, Rectifie the Globe to the Latitude of your Place

Place, and bring the Place of the Sun to the Eaſt ſide the Ho-
rizon; and the Stars then cut by the Eaſtern Semi-Circle of the
Horizon, Riſe *(Coſmically* ; and thoſe Stars cut by the Weſtern
Semi-Circle of the Horizon, Set *(Coſmically.*

Example.

Novemb. 9. I would know what Stars Riſe and Set *(Coſmically.* here at *London.* The Suns Place found, as by the third
Probleme is ♏ 27. Therefore I bring ♏ 27. to the Eaſt ſide the
Horizon, and in the Eaſtern Semi-Circle I find Riſing with the
Sun the right *Wing* of *Cygnus,* the Star in the end of *A-
quila's tail, Serpentarius* and *Centaurus* : Therefore theſe Conſtellations are ſaid to rise *(Coſmically.* In the Weſtern Semi-Circle
of the Horizon I find Setting *Andromeda,* the *Triangle, Taurus, Orion, Canis Major,* and *Argo Navis* ; Therefore I ſay,
theſe Conſtellations Set *(Coſmically.*

PROB. XXXVI.

To find by the Coeleſtial Globe the Acronical *Riſing and
Setting of the Stars.*

THe Stars that Riſe when the Sun Sets, are ſaid to Riſe *A-
cronically.* And,
 The Stars that Set with the Sun, are ſaid to Set *Acronically.*
 To find theſe, Rectifie the Globe to the Latitude of your
Place, and bring the Place of the Sun to the Weſt ſide the Hori-
zon; and the Stars then cut by the Eaſtern Semi-Circle of the
Horizon, Riſe *Acronically* : And thoſe Stars cut by the Weſtern
Semi-Circle of the Horizon, Set *Acronically.*

Example.

November 9. I would know what Stars Riſe and Set *Acronically* here at *London.* The Suns Place as before, is ♏ 27.
Therefore I bring ♏ 27. to the Weſt ſide the Horizon ; and in
the Eaſtern Semi-Circle I find Riſing the *Southern Fiſh, Fomahant, Cetus, Taurus, Auriga,* and the *Feather* in *Caſtor's Cap.*
Therefore theſe Conſtellations are ſaid to Riſe *Acronically.* In

L 3   the

the Weftern Semi-Cirde of the Horizon I find Setting the *Lyons tail, Virgo, Scorpio, and Sagittarius,* Therefore I fay, thefe Conftellations Set *Acronically.*

PROB. XXXVII.

To find by the Coeleftial Globe the Heliacal Rifing, *and* Setting *of the Stars.*

WHen a Star formerly in the Suns Beams gets out of the Suns Beams it is faid to *Rife Heliacally.* And.

When a Star formerly out of the Suns Beams, gets into the Suns Beams, it is faid to *fet Heliacally.*

A Star is faid to be in the Suns Beams, when it is made inconfpicuous by reafon of its neernefs to the Suns Light. The Bigger Stars are difcernable more neer the Suns Light, then the Leffer are : For, Stars of the firft Magnitude may (according to the received Rules of ancient Authors) be feen when the Sun is but 12. degrees below the Horizon but Stars of Second Magnitude cannot be feen unlefs the Sun be 13. degrees below the Horizon : Stars of the third Magnitude require the Sun to be 14. degrees below the Horizon ere they can be feen; of the fourth Magnitude 15. degrees. of the fifth Magnitude 16. degrees of the fixth Magnitude 17 degrees ; the Nebulous ones 18. degrees. Yet this Rule is not fo certain but that either dear or cloudy weather may alter it. Read more of this fubject in M*r* *Palmer* on the *Planifphear.* Book 4. Chap. 20

Now to find the Time that any Star fhall Rife Heliacally. Do thus Rectifie the Globe and Quadrant of Altitude to your Latitude. Then bring the given Star to the Eaft fide the Horizon, and turn the Quadrant of Altitude into the Weft fide, and fee what degree of the *Ecliptick* is elevated fo many degrees above the Horizon as the Magnitude of the Star you enquire after requires, according to the foregoing Rules; for the oppofite degree of the *Ecliptick* is the degree the Sun fhall be in when that Star Rifes *Heliacally.* Having the degree of the *Ecliptick* the Sun is in, you may find the Day of the Moneth, by the 4th Probleme.

Exam-

Example.

I would know when *Cor Leonis* fhall Rife *Heliacally* here at *London* : Therefore I Rectifie the Globe and Quadrant of Altitude for *London,* and bring *Cor Leonis* to the Eaft fide the Horizon, and turn the Quadrant of Altitude into the Weft ; and becaufe *Cor Leonis* is a Star of the firft Magnitude, therefore I fee what degree of the *Ecliptick* is elevated in the Weft fide the Horizon 12. degrees on the Quadrant of Altitude, and find ♓ 9. deg. Now the degree of the *Ecliptick* oppofite to ♓ 9. is ♍ 9. Therefore I fay, when the Sun comes to ♍ 9. degrees (which by the 4th Probleme I find is *Auguft.* 23.) *Cor Leonis* fhall *Rife Heliacally.*

For the *Heliacal Setting.*

The Globe, &c. Rectified, as before : Bring the Star to the Weft fide the Horizon, Then fee what degree of the *Ecliptick* is elevated on the Quadrant of Altitude fo many degrees as the Stars Magnitude requires; for when the Sun comes to the oppofite degree of the *Ecliptick* that Star fhall *fet Heliacally.*

Example.

I would know when *Bilanx* a Star in the *Beam* of the *Scales,* will *Set Heliacally* here at *London.* The Globe and Quadrant Rectified, I bring *Bilanx* to the Weft fide the Horizon, and turn the Quadrant of Altitude into the Eaft ; Then I examine what degree of the *Ecliptick* is elevated 13. degrees of the Quadrant of Altitude (becaufe *Bilanx* is a Star of the fecond Magnitude) and find ♉ $4\frac{1}{2}$, oppofite to ♉ $4\frac{1}{2}$. is ♏ $4\frac{1}{2}$. Therefore I fay, When the Sun comes to ♏ $4\frac{1}{2}$. (which by Probleme 4. will be *October* 18) *Bilanx* fhall fet *Heliacally.*

PROB. XXXVIII.

To find the Diurnal and Nocturnal Arch of the Sun, or Stars, in any given Latitude.

THe Semi-Diurnal Arch is the number of degrees of the Equator that paffes through the Meridian whiles the Sun or
any

any Star is aſcending above the Eaſt ſide the Horizon to the Meridian. To know the number of degrees it contains, Rectifie the
Globe to the given Latitude, and bring the Place of the Sun or
Star to the Eaſt ſide the Horizon, and note what number of degrees of the Equinoctial is then cut by the Meridian : Then remove the Place of the Sun or Star to the Meridian, and ſee
again what number of degrees of the Equinoctial is then
cut by the Meridian, and ſubſtract the former from the
latter, and the remainder ſhall be the number of degrees of the
Sun or Stars Semi-Diurnal Arch. But Note, If the Equinoctial
point ♈ paſs through the Meridian while the Sun or Star
is turned from the Eaſt ſide the Horizon to the Meridian, then
you muſt ſubſtract the number of degrees of the Equinoctial
cut by the Meridian when the Sun or Star is at the Eaſt ſide the
Horizon from 360. degrees, and to the remainder ad the number of degrees of the Equinoctial that comes to the Meridian
with the Place of the Sun or Star, and the Sum of them both is
the number of degrees of the Sun or Stars Semi-diurnal Arch;
which being doubled is the number of degrees of the whole Diurnal Arch : and which being ſubſtracted from 360, gives
the Nocturnal Arch.

Example, of the Sun.

Having Rectified the Globe, I would *May* 10. know the
Diurnal Arch of the Sun : His Place found by *Prob.* 3. is ♉
29. Therefore I bring ♉ 29. to the Faſt ſide the Horizon, and
find then at the Meridian 299. degrees 30. min. of the Equinoctial ; then I turn the Place of the Sun to the Meridian, and
find 56. deg. 30. min. of the Equinoctial come to the Meridian
with it. Here the Equinoctial point ♈ paſſes through the Meridian while the Sun moves between the Horizon and the Meridian ; Therefore as aforeſaid, I ſubſtract the firſt number of
degrees and minutes *viz.* 299. deg. 30. min. from 360. degrees,
and there remains 60. degr. 30. min. for the number of degrees
and minutes contained between the degree of the Equinoctial at
the Meridian and the Equinoctial point ♈ ; and to this 60. deg.
30. min. I ad the ſecond number of degrees and minutes, *viz.*
56. deg. 30. min. the number of degrees and minutes between
the point ♈ and the deg. of the Equinoctial at the Meridian, and
they make together 117. degrees, for the Suns Semi diurnal
Arch;

Arch; By doubling of which, you have 234. degrees, for the Suns Diurnal Arch : And by fubftracting 234. (the Diurnal Arch) from 360. you have 126. degrees, for the Suns Nocturnal Arch.

Example, for a Star.

I take *Sirius,* a bright Star in the *Great Dogs mouth.* The Globe rectified, as before; I bring *Sirius* to the Eaft fide the Horizon, and find then 29. degrees 30 minutes of the Equinoctial at the Meridian, then I turn *Sirius* to the Meridian and find 97. degrees 38 minutes of the Equinoctial come to the Meridian with it : Therefore I fubftract the firft number *viz.* 29. degrees 30. minutes, from the fecond, 97. 38, and the remains is 68. degrees 8 minutes, for the Semi-diurnal Arch of *Sirius.*

His Nocturnal Arch you may find as before.

PROB. XXXIX.

To find the Azimuth *and* Almicantar *of any Star.*

THis Probleme is like the 22, and 23. Problemes, which fhew the finding the *Azimuth* and *Almicantar* of the Sun ; only, whereas there you were directed to bring the degree of the Sun to the Quadrant of Altitude, you muft now bring the Star propofed to the Quadrant of Altitude ; and by the Directions in thofe Problemes the refolution will be found.

PROB. XL.

To find the Hour *of the* Night, *by obferving two known Stars in one* Azimuth, *or* Almicantar.

REctifie the Globe Quadrant and Hour Index. Then find the two known Stars on the Globe ; and if the two Stars be in one *Azimuth,* turn about the Globe and Quadrant of Altitude till you can fit the two Stars to ly under the graduated edge of the Quadrant of Altitude : fo fhall the Index of the Hour-Circle point at the Hour of the Night. If

M the

the two Stars be in one *Almicantar*, Turn the Globe forward
or backward till the two Stars come to fuch a Pofition that by
moving the Quadrant of Altitude, the fame degree on it will ly
on both the Stars ; fo fhall the Index of the Hour-Circle point
at the Hour of the Night.

PROB. XLI.

The Hour *given that any Star in Heaven comes to the Me-*
ridian, to know thereby the Place of the Sun, and by con-
fequence the Day of the Moneth, though it were loft.

BRing the Star propofed to the Meridian, and turn the In
dex of the Hour-Circle to the Hour given, Then turn a-
bout the Globe till the Index point at the Hour of 12,
for Noon ; and the Place of the Sun in the *Ecliptick*
fhall be cut by the Meridian.

Example.

March 7. at 11. aclock at Night the *Pointers* come to the
Meridian of *London.* Therefore I place the *Pointers* on the *Cae-*
leftial Globe under the Meridian, and turn the Index of the
Hour-Circle to 11. paft Noon, afterwards I turn back the Globe
till the Index point to 12. at Noon ; Then looking in the *Eclip-*
tick, I find the Meridian cuts it in ♓ 26. 45. minutes; Therefore
I fay, when the *Pointers* come to the Meridian at 11. a clock at
Night, the Place of the Sun is ♓ 26. 45. Having thus the Place
of the Sun, I may find the Day of the Moneth by the fourth
Probleme ; and fo either know the Day that the *Pointers* come
to the Meridian at 11. a clock at Night, or at any other Hour
given.

The Day of the Moneth might alfo be found by the Declination and the Quarter of the *Ecliptick* the Sun is in, given : For
the Meridian will cut the degree of the Suns Place in the *Eclip-*
tick in the Parallel of Declination : So that having refpect to the
Quarter of the *Ecliptick,* you'le find the Suns Place ; and having
the Suns Place, you may as aforefaid find the Day of the
Moneth.

PROB.

PROB. XLII.

The Day of the Moneth given, to find in the Circle of Letters on the Plain of the Horizon, the Day of the Week.

THe feven Daies of the Week were by the Idolatry of the ancient Roman Heathenifh Times Dedicated to the Honour of feven of their Gods, which we call *Planets.* The firft is the moft eminent, and therefore doubtlefs by them fet in the firft Place, called *Dia Solis,* or the Suns Day : The fecond *Dia Luna,* the Moons Day : The third *Dia Martis,* the Day of *Mars:* by us called *Tuefday* : The fourth *Dia Mercurius, Mercuries* Day : by us called *Wednefday* ; from *Woden,* an Idol the *Saxons* Worfhipt, to whofe Honour they Dedicated that Day, and is by all thofe *Germain* Nations ftill called *Wodenfdagh* : The fifth *Dia Jovis, Jupiter* or *Joves* Day : which doubtlefs the *Saxons* (from whom probably we receive it) called *Donder-dagh,* becaufe *Jupiter* is the God of Thunder ; and we either by corruption or for fhortnefs, or both, call it *Thurfday:* The fixth *Dia Veneris,* the Day of *Venus* : but the *Saxons* transferring her Honour to another of their Goddeffes named *Fria,* called it *Fridagh* : and we from them call it *Fryday* : The feventh is *Dia Saturnis, Saturns* Day.

The fame Day of the Moneth in other Years happens not on the fame Day of the Week, therefore the Dominical Letter for one Year is not the fame it is the next : Now becaufe you cannot come to the knowledge of the Day of the Week unlefs you firft know the Sundaies Letter, therefore have I in Prob. 53 inferted a Table of M*r Palmers,* by which you may find the Dominical or Sundaies Letter for ever; and having the Dominical Letter you may in the Circle of Letters on the Horizon find it neer the day of that Moneth, and count that for *Sunday,* the next under it for *Monday,* the next under that for *Tuefday,* and fo in order, till you come to the Day of the Moneth.

Example.

I would know what Day of the Week *June* 1. *Anno* 1658. Old Style, falls on ; I find by the Table aforefaid the Dominical

Letter

Letter is C, then I look in the Calender of Old Style for *June* 1. and againſt it I find Letter E, which becauſe it is the ſecond Letter in order from C, therefore it is the ſecond Day in order from *Sunday,* which is *Tueſday.*

PROB. XLIII.

The Azimuth *of any Star given, to find its* Hour *in any given* Latitude.

THe Hour of a Star is the number of Hours that a Star is diſtant from the Meridian. To find which, Rectifie the Globe and Quadrant of Altitude, and bring the Star propoſed to the Meridian, and the Index of the Hour-Circle to 12. Then place the lower end of the Quadrant of Altitude to the given *A-zimuth* in the Horizon, and turn the Globe till the Star come to the graduated edge of the Quadrant of Altitude ; ſo ſhall the Index of the Hour-Circle point at the Hour of the Star. Only this caution you muſt take; If the Star were turned from the Meridian towards the Eaſtern ſide of the Horizon, you muſt ſubſtract the number of Hours the Index points at from 12. and the remainder ſhall be the Hour of the Star. But if the Star were turned from the Meridian towards the Weſt ſide the Horizon, the Hour the Index points at is (without more adoe) the Hour of the Star.

PROB. XLIV.

How you may learn to know all the Stars in Heaven, by the Coeleſtial Globe.

REctifie the Globe, Quadrant, Hour-Index and Horizon, as by Prob. 2. Then turn about the Globe till the Index of the Hour-Circle point at the Hour of the Night on the Hour-Circle. Then if every Star on the Globe had a hole in the midſt, and your Ey were placed in the Center of the Globe; you might by keeping your Ey in the Center and looking through any Star on the Globe ſee its Match in Heaven : that is, the ſame Star in Heaven which that Star on the Globe repreſents: for from the Center of the Globe there proceeds a ſtraight
line

line through the Star on the Globe, even to the fame Star in Heaven. Therefore thofe Stars that are in the *Zenith* in Heaven,
will then be in the *Zenith* on the Globe ; thofe that are in the
Eaft in Heaven, will be in the Eaft on the Globe ; thofe in the
Weft in Heaven, in the Weft on the Globe ; and thofe Stars
that are in any Altitude in Heaven, will at the fame time have
the fame Altitude on the Globe ; So that if you fee any Star in
Heaven whofe Name you defire to know, you need but obferve
its *Azimuth* and Altitude, and in the fame *Azimuth* and Altitude on the Globe, you may find the fame Star : and if it be an
eminent Star, you will find its Name adjoyned to it.

Example.

December 10. at half an hour paft 9. a clock at Night, here at
London, I fee two bright Stars at a pretty diftance one from another in the South; I defire to know the Names of them; Therefore having the Globe rectified to the Latitude of *London,* and
the Quadrant of Altitude fcrewed to the *Zenith,* the Hour-Index
alfo Rectified, and the Horizon pofited Horizontally, as by Prob.
2. I obferve the Altitude of thofe Stars in Heaven, (either with a
Quadrant, Aftrolabe, Crofs-ftaff, or the Globe it felf, as hath
been fhewed Prob. 13, 16.) to be, the one 78. degrees, the o-
ther 42. degrees above the Horizon. Therefore having their
Altitudes, I count the fame number of degrees as for the firft 78.
upon the Quadrant of Altitude upwards, and turn it into the
South, under the Meridian, and fee what Star is under 78. degrees, for that is the fame Star on the Globe which I faw in
Heaven. Now at the firft examination of the Globe you may
fee that that Star is placed in the Ey of that Afterifme which is
called *Caput Medufa,* and indeed, that being the only Star of
Note in that Conftellation, bears the Name of the whole Con-
ftellation. The other Stars about it you may eafily know by their
Scituation. As, Seeing two little Stars to the Weftwards of that
Star in Heaven, you may fee on the Globe that the hithermoft is
in the other Ey of *Caput Medufa,* and the furthermoft in the
Hair or *Snakes* of the fame Afterifme. Looking a little to the
Southwards of thofe Stars in Heaven, you may fee two other fmal
Stars a little below thofe in the *Eyes* ; Therefore to know thofe
alfo, you may look on the Globe, and fee that there is one on the

Noſe, and another Starre in the *Cheek* of *Caput Meduſa*

In like manner for the ſecond Star in the Meridian, which is 42 degrees above the Horizon : If you move the Quadrant of Altitude (as before) to the South or Meridian, and count 42 degrees upon the Quadrant of Altitude, you will find a Star of the ſecond Magnitude in the *Mouth* of the *Whale* : Therefore you may ſay, that Star in Heaven is in the *Mouth* of the *Whale* : and becauſe cloſe to it on the Globe is written *Menkar,* Therefore you may know the name of that Star in Heaven is *Menkar.*

In the South Eaſt and by South 56 degrees above the Horizon, I ſee a very bright Star in Heaven ; therefore I bring the Quadrant of Altitude to the South Eaſt and by South point in the Horizon, and find under 56 degrees of the Quadrant of Altitude a great Star, to which is prefixed the name *Occulus Taurus;* Therefore I ſay, the name of that Star in Heaven is *Occulus Taurus.*

In the South Eaſt in Heaven you may ſee three bright Stars ly directly in a ſtraight line from one another, the middlemoſt whereof is 25. degrees or thereabouts above the Horizon, therefore bring the Quadrant of Altitude to the South Eaſt point of the Horizon, and about 25 degrees above the Horizon you will ſee the ſame great Stars on the Globe, in the *Girdle* of *Orion:* Therefore thoſe Stars are called *Orions Girdle.*

At the ſame time South Eaſt and by Eaſt you have about 10 degrees above the Horizon the brighteſt Star in Heaven, called *Sirius,* in the *Mouth* of the *Great Dog* ; *Canicula* a bright Star in the *Little Dog* Eaſt and by South, about 25 degrees above the Horizon: *Cor Leonis* juſt Riſing Eaſt North Eaſt : you have alſo at the ſame time on the Eaſt ſide the Horizon, the *Twins,* *Auriga,* the *Great Bear* ; and divers other Stars, eminent both for their ſplendor and Magnitude.

In the Weſt ſide the Horizon you have South Weſt and by Weſt about 4 degrees above the Horizon a bright Star in the *Right Leg* of *Aquarius* : and all along to the Southwards in *Cetus* the *Whale,* you have other eminent bright Stars : More upwards towards the *Zenith* you have a bright Star in the *Line* of the two *Fiſhes* : Higher yet, you have the firſt Star in ♈, an eminent Star, becauſe the firſt in all Catalogues that we have cognizance of ; and therefore probably in the Equinoctial Colure when the Stars were firſt reduced into Conſtellations : yet more

neer

neer the *Zenith* you have a bright Star in the *Left Leg* of *Andromeda* : From thence towards the North, you find other very eminent bright Stars in *Caſſiopea, Cepheus, Vrſa Minor,* in the Tail whereof is the *Pole Star* : and *Draco* : *Hecules* : where you turn back, to *Lyra, Cygnus, Pegaſus,* the *Dolphin,* &c. all which, or any other, you may eaſily know by their Altitude above the Horizon, and the point of the Compaſs they bear upon.

Thus knowing ſome of the moſt eminent Fixed Stars, you may by the Figure of the reſt come to the knowledge of them alſo. For Example, Looking towards the North North Eaſt in Heaven, you may ſee ſeven bright Stars conſtituted in this Figure; Therefore looking towards the ſame Quarter on the Globe, you may (without taking their Altitude) ſee the ſame Stars lying in the ſame Figure in the hinder parts of the *Great Bear* ; from whence you may conclude, that thoſe Stars in Heaven are ſcituate in the hinder parts of the Aſteriſme called *Vrſa Major.*

Yet nevertheleſs you may ſee ſome Stars of Note in Heaven, which you ſhall not find on the Globe, and thoſe in or neer about the Ecliptick : They are called *Planets,* and cannot be placed on the Globe, unleſs it be for a particular Time, with Black Lead, or ſome ſuch thing that may be rubbed out again : Becauſe they having a continual motion alwaies alter their Places. Of thoſe there are five in number, beſides the *Sun* and *Moon,* which are alſo Planets, though they ſhew not like Stars. Theſe five are called *Saturn, Jupiter, Mars, Venus, Mercury* ; yet *Mecury* is very rarely ſeen: becauſe he never Riſing above an Hour before the Sun, or Setting above a Hour after, for the moſt part hath his light ſo overſpread with the dazelling Beams of the glittering Sun, that ſometimes when he is ſeen he ſeems rather to be a Mote in the Suns Beams, then a Body endowed with ſo much brightneſs as Stars and Planets ſeem to be.

Now there are divers waies (by ſome of which you may at all times) know thoſe Planets from the Fixed Stars : as firſt, Their not twinkling, for therein they differ from fixed Stars ; becauſe they moſt commonly do twinkle, but Planets never ; unleſs it be ♂ *Mars* ; and yet he twinkles but very ſeldom neither.

Secondly, They appear of a conſiderable Magnitude, as ♃ ſome

ſometimes appears greater ly far then a Star of the firſt Magnitude; and ☿ many times bigger then he. They are both glittering Stars, of a bright Silver collure; but ♀ moſt radient, eſpecially
when ſhe is in her *Perigeon.* ♂ appears like a Star of the ſecond
Magnitude ; and is of a Copperiſh colloure. ♄ ſhewes like a
Star of the third Magnitude, and is of a Leaden Collour; and he
(of all the others,) is moſt difficult to be known from a fixed Star;
partly becauſe of his minority, and partly becauſe of the ſlowneſs
of his motion. ☿ is very ſeldom ſeen (as aforeſaid) unleſs it be in
a Morning when he Riſes before the Sun, or in an Evening when
he Sets after the Sun : He is of a Pale Whitiſh Collour, like
Quick ſilver, and appears like a Star of the third Magnitude. He
may be known by the Company he keeps, for he is never above
29. degrees diſtant from the Sun.

Thirdly, The Planets may be known from fixed Stars by their
Azimuths and Altitudes obſerved: (as hath been taught before) for
if when you have taken the Azimuth and Altitude of the Star in
Heaven you doubt to be a Planet, and you find not on the Globe
in the ſame Azimuth and Altitude a Star appearing to be of the
ſame Magnitude that that in Heaven appears to be, you may
conclude that that in Heaven is a Planet. Yet notwithſtanding
it may happen that a Planet may be in the ſame degree of
Longitude and Latitude in the Zodiack that ſome eminent
fixed Star is in; as in the degree and minute of Longitude and Latitude that *Cor Leonis,* or the *Bulls Ey,* or *Scorpions heart* is in,
and ſo may eclipſe that Star, by being placed between us and it :
But that happens very ſeldom and rarely; but if you doubt it, you
may apply your ſelf to ſome other of the precedent and ſubſe-
quent Rules here ſet down for knowing Planets from fixed Stars.

The fourth way is by ſhifting their Places ; for the Planets
having a continual motion, do continually alter their Places : as
♂ moves about half a degree in a day : ♀ a whole degree;
but ♃ and ♄ move very ſlowly ; ♃ not moving above 5. minutes, and ♄ ſeldom above 2. minutes. Yet by their motions alone
the Planets may be known to be Planets, if you will preciſely obſerve their diſtance from any known fixed Star in or near the E-
cliptick as on this Night, and the next Night after obſerve whether
they retain the ſame diſtance they had the Night before; which if
they do, then are they fixed Stars ; but if they do not then are
they Planets: yet this Caution is to be given you in this Rule alſo,
                                                              That

That the Planets fometimes are faid to be *Stationary,* as not al-
tering 1. minute in Place, forwards, or backwards in 6. or 7.
daies together. Therefore, if you find caufe to doubt whether
your Star be a Planet, or a fixed Star, you may for the help of
your underftanding confer with fome of the former Rules, unlefs
you are willing to wait 8 or 9 daies longer, and fo by obfervation
of its motion refolve your felf. Or,

Fifthly, you may apply your felf to an *Ephemeris* for that
Year, and fee if on that day you find any Planet in the degree and
minute of the Zodiack you fee the Star you queftion in Heaven ;
and if there be no Planet in that degree of the Zodiack, you may
conclude it is no Planet, but a fixed Star.

PROB. XLV.

How to hang the Terreftrial *Globe in fuch a pofition that by
the Suns fhining upon it you may with great delight at
once behold the demonftration of many* Principles *in A-
ftronomy, and* Geography.

TAke the Terreftrial Ball out of the Horizon, and faften a
thred on the Brazen Meridian to the degree of the Latitude of your Place; by this thred hang the Globe in
a place where the Suns Beams may have a free accefs
to it ; Then direct the Poles of the Globe to their proper Poles
in Heaven, the North Pole to the North, and the South Pole to
the South ; and with a thred faftned to either Pole, brace the
Globe, fo, that it do not turn from his pofition : then bring
your Habitation to the Meridian ; fo fhall your Terreftrial
Globe be Rectified to correfpond in all refpects with the Earth
it felf ; even as in Prob. 44. the Celeftial Globe doth ; the
Poles of the Globe, to the Poles of the World ; the Meridian of
the Globe, to the Meridian of the World ; and the feveral Regions on the Globe made Correfpondent to the fame Regions on
the Earth : So that with great delight you may behold,

1.   How the counterfeit Earth (like the true one) will have
one Hemifphear Sun fhine light, and the other fhadowed, and as
it were dark. By the fhining Hemifphear you may fee that it is
Day in all Places that are fcituate under it; for on them the Sun
doth fhine; and that it is Night at the fame time in thofe Places

                              N   that

that are ſituate in the ſhadowed Hemiſphear ; for on them the Sun doth not ſhine ; and therefore they remain in darkneſs.

2. If in the middle of the enlightned Hemiſphear you ſet a Spherick Gnomon Perpendicularly, it will project no ſhadow, but ſhews that the Sun is juſt in the Zenith of that Place ; that is, directly over the heads of the Inhabitants of that Place : and the point that the Spherick Gnomon ſtands on, being removed to the Meridian, ſhews the Declination of the Sun on the Meridian for that Day.

3. If you draw a Meridian line from one Pole to the other, in all Places under that line, it is Noon : in thoſe Places ſcituate to the Weſt, it is Morning ; for with them the Sun is Eaſt : and in thoſe Places ſcituate to the Eaſt, it is Evening; for with them the Sun is Weſt.

4. Note the degree of the Equator where the enlightned Hemiſphear is parted from the ſhadowed ; for the number of degrees of the Equator intercepted between that degree and the Meridian of any Place, converted into Hours (by accounting for every 15. degrees 1. Hour) ſhews, if the Sun be Eaſtwards of that Place, how long it will be ere the Sun Riſes, Sets, or comes to the Meridian of that Place : or if the Sun be Weſtward of that Place, how long it is ſince the Sun Roſe, or Set, or was at the Meridian of that Place.

5. The Inhabitants of all Places between the enlightned and ſhadowed Hemiſphear, behold the Sun in the Horizon : Thoſe Weſtwards of the Meridian Semi-Circle drawn through the middle of the enlightned Hemiſphear behold the Sun Riſing : Thoſe in the Eaſt, ſee it Setting.

6. So many degrees as the Sun reaches beyond either the North or South Pole, ſo many degrees is the Declination of the Sun, either Northwards or Southwards : and in all thoſe Places comprehended in a Circle deſcribed at the termination of the Sun-ſhine, about that Pole, it is alwaies Day, till the Sun decreaſe in Declination : for the Sun goes not below their Horizon : as you may ſee by turning the Globe about upon its Axis : and in the oppoſite Pole at the ſame diſtance, the Sun-ſhine not reaching thither, it will be alwaies Night, till the Sun decreaſe in Declination : becauſe the Sun Riſes not above their Horizon.

7. If you let the Globe hang ſteddy, you may ſee on the Eaſt ſide of the Globe, in what Places it grows Night ; and on the
Weſt

Weſt ſide the Globe how by little and little the Sun encroaches upon it; and therefore there makes it Day.

8. If you make of Paper or Parchment a narrow Girdle, to begirt the Globe juſt in the Equinoctial, and divide it into 24. e-qual parts, to repreſent the 24. hours of Day and Night, and mark it in order with I, II, III, &c. to XII. and then be-gin again with I, II, III, &c. to the other XII. you may by placing one of the XII^r. upon the Equinoctial under the Meridian of your Place, have a continual Sun-Dyal of it, and the hour of the Day given on it, at once in two places; one by the parting the enlightned Hemiſphear from the ſhadowed on the Eaſtern ſide, the other by the parting the enlightned Hemiſphear from the ſhadowed on the Weſtern ſide the Globe. Much more might be ſaid on this Probleme: But the Ingenuous Artiſt may of himſelf find out diverſities of Speculations: therefore I forbear.

PROB. XLVI.

To know by the Terreſtrial Globe in the Zenith of what Place of the Earth the Sun is.

THis may be performed by the former Probleme in the Day time, if the Sun ſhines: but not elſe. But to find it at all times, do thus. Bring the Place of your Habitation to the Meri-dian, and the Index of the Hour-Circle to 12; Then turn the Globe Eaſtwards, if Afternoon, or Weſtwards, if Before Noon, till the Index of the Hour-Circle paſs by ſo many Hours from 12. as your Time given is, either before or After-Noon: ſo ſhall the Sun be in the Zenith of that Place where the Meridian interſects the Parallel of the Suns Declination for that Day.

Example.

May 10 at ¾ of an hour paſt 4. a clock After Noon. I would know in what Place of the Earth the Sun is in the Ze-nith. My Habitation is *London.* Therefore I bring *London* to the Meridian, and the Index of the Hour-Circle to 12. and becauſe it is After Noon: I turn the Globe Eaſtwards, till the Index paſſes through 4 hours and 3 quarters, or (which is all one) till 70 degrees 15 minutes of the Equator paſs through the Me-

ridian.

ridian. Then I find by Prob. 5. the Suns Declination is 20. degrees 5. minutes which I find upon the Meridian, and in that
Place juſt under that degree and minute on the Globe, the Sun
is in the *Zenith:* which in this Example is in the North Eaſt
Cape of *Hiſpaniola.*

Having thus found in what Place of the Earth the Sun is in
the *Zenith.* Bring that Place to the Meridian, and Elevate its
reſpective Pole according to its reſpective Elevation; ſo ſhall all
Places cut by the Horizon have the Sun in their Horizon: Thoſe
to the Eaſtwards ſhall have the Sun Setting ; thoſe to the
Weſtward ſhall have it Riſing in their Horizon : thoſe at the
Interſection of the Meridian and Horizon under the Elevated
Pole, have the Sun in their Horizon at loweſt, but Riſing ; thoſe
at the Interſection of the Meridian and Horizon under the
Depreſſed Pole, have the Sun in their Horizon at higheſt, but
Setting. Thus in thoſe Countries that are above the Horizon it
is Day-light, and in thoſe but 18 degrees below the Horizon, it is
Twilight : But in thoſe Countries further below the Horizon
it is at that time dark Night : And thoſe Countries within the
Parallel of the ſame number of degrees from the Elevated Pole
that the Suns Declination is from the Equinoctial, have the Sun
alwaies above the Horizon, till the Sun have leſs Reſpective
Declination then the Elevated Pole; and thoſe within the ſame
Parallel of the Depreſſed Pole have the Sun always below
their Horizon, till the Sun inclines more towards the Depreſſed
Pole ; As you may ſee by turning about the Globe ; for in this
poſition, that portion of the Globe intercepted between the Elevated Pole, and the Parallel Circle of 20. degrees 5. minutes
from the Pole doth not deſcend below the Horizon: neither doth
that portion of the Globe intercepted between the Depreſſed
Pole and the Parallel Circle within 20. degrees 5. minutes of
that Pole, aſcend above the Horizon.

PROB. XLVII.

To find in what different Places of the Earth the Sun hath
  the ſame Altitude, *at the ſame time.*

FInd by the former Probleme in what Place of the Earth
the Sun is in the *Zenith,* and bring that Place on the
                                                        Globe

Globe to the *Zenith*, and on the Meridian [there] screw
the Quadrant of Altitude, and turn it about the Horizon, de-
scribing degrees of *Almicantars* thereby, as by Prob. 23. and all
those Countries in any *Almicantar* on the Globe shall have the
Sun Elevated the same number of degrees above their Horizon.
Thus those Countries in the tenth *Almicantar* shall have the Sun
Elevated 10. degrees above their Horizon ; those in the 20th
Almicantar shall have the Sun Elevated 20 degrees above their
Horizon ; those in the 30th, 30. degrees &c. So that you may
see, when the Sun is in the *Zenith* of any Place, All the Countries
or Cities in any *Almicantar* have the Sun in one heighth at the
same time above their Horizon. But to find in what different
Places the Sun hath the same heighth at the same time, as well
Before or After Noon, as at Full Noon ; and that in Countries
that have greater Latitude then the Suns greatest Declination,
(and therefore cannot have the Sun in their *Zenith*,) requires
another Operation.

Therefore, Elevate its respective Pole according to your respective Latitude ; and let the Degree of the Brazen Meridian
which is in the *Zenith* represent your Habitation, and the degree
of the Ecliptick the Sun is in represent the Sun : Then bring the
Sun to the Meridian, and the Index of the Hour-Circle to 12,
and turn the Globe Eastwards, if Before Noon, or Westwards,
if After Noon, till the Index point to the Hour of the Day. Then
place the lower end of the Quadrant of Altitude to the East
point of the Horizon, and move the upper end (by sliding the
Nut over the Meridian) till the edge of the Quadrant touch the
place of the Sun : Then see at what degree of the Meridian the
upper end of the Quadrant of Altitude touches the Meridian and
substract that number of Degrees from the Latitude of your
Place, and count the number of remaining degrees on the Meridian, on the contrary side the degree of the Meridian where the upper end of the Quadrant of Altitude touches the Meridian, and
where that number of degrees ends on the Meridian, in that Latitude and your Habitations Longitude, hath the Sun the same
heighth at the same time.

Example.

May 10. at 53. minutes past 8. a clock in the Morning I

would

would know in what Place the Sun ſhall have the ſame Altitude it ſhall have at *London. London's* Latitude found by *Prob.*
1. is 51½ degrees Northwards : And becauſe the Elevation of
the Pole is equal to the Latitude of the Place (as was ſhewed
Prob. 15.) Therefore I Elevate the North Pole 51½ degrees, ſo
ſhall 51½ degrees on the Meridian be in the *Zenith* : This 51½
degrees on the Meridian repreſents *London.* The Suns Place found
by *Prob.* 3. is ♉ 29. Therefore I bring ♉ 29 to the Meridian, and
the Hour Index to 12. on the Hour Circle: Then I turn the Globe
Eaſtwards (becauſe it is before Noon) till the Index point at 8.
hours 53 minutes on the Hour-Circle, and place the lower end of
the Quadrant of Altitude to the Eaſt point in the Horizon, and
ſlide the upper end either North or Southwards on the Meridian
till the graduated edge cut the degree of the Ecliptick the Sun is
in : Then I examine on the Meridian what degree the upper end of the Quadrant of Altitude touches ; which in this
example, I find is 38½ degrees. Therefore I ſubſtract 38½
from 51½ *Londons* Latitude, and there remains 13. Then
counting on the Meridian 13. degrees backwards, from the
Place where the Quadrant of Altitude touched the Meridian, I
come to 25½ on the Meridian, Northwards. Therefore I ſay, In
the North Latitude of 25½ degrees, and in the Longitude of *London* (which is in *Africa,* in the Kingdom of *Numidia*) the Sun
May 10. at 53. minutes paſt 8. a clock in the Morning hath the
ſame Altitude above the Horizon it hath here at *London.*

 The Quadrant of Altitude thus applyed to the Eaſt point of
the Horizon makes right angles with all points on the Meridian,
even as all the Meridians proceeding from the Pole, do with the
Equator : therefore the Quadrant being applyed both to the
Eaſt point, and the Suns Place, projects a line to interſect the Meridian Perpendicularly in equal degrees; from which interſection
the Sun hath at the ſame time equal Heighth, be the degrees
few or many ; for thoſe 5. degrees to the Northwards of this interſection, have the Sun in the ſame heighth that they 5 degrees
to the Southwards have it : and thoſe 10, 20, 30. degrees,
more, or leſs, to the Northwards, have the Sun in the ſame
heighth that they have that are 10. 20. 30. degrees more or leſs
to the Southwards : So that this *Prob.* may be performed another way more eaſily, with your Compaſſes, Thus : Having firſt
rectified the Globe, and Hour Index, Turn about the Globe till
                                                      the

the Hour Index point to the Hour of the Day ; Then pitch one
foot of your Compaſſes in the Suns Place, and extend the other
to the degree of Latitude on the Meridian, which in this example is 51½ degrees North ; then keeping the firſt foot of your
Compaſſes on the degree of the Sun, turn about the other foot to
the Meridian, and it will fall upon 25½. as before.

Blaew commenting upon this Probleme, takes notice how
groſly they ere that think they can find the heighth of the Pole
at any Hour of the Day, by the Suns height : becauſe they
do not conſider that it is impoſſible to find the Hour of the Day,
unleſs they firſt know the height of the Pole.

PROB. XLVIII.

*To find the length of the Longeſt and Shorteſt Artificial
Day or Night.*

THe Artificial Day is that ſpace of Time which the Sun
is above the Horizon of any Place : and the Artificial Night is that ſpace of Time which the Sun is under
the Horizon of any Place. They are meaſured in the
Hour Circle, by Hours and Minutes.

There is a conſtant unequallity of proportion in the Length of
theſe Daies and Nights ; which is cauſed both by the alteration
of the Suns Declination, and the difference of the Poles Elevation.

Thoſe that inhabit on the North ſide the Equator have their
longeſt Day when the Sun enters ♋ ; and thoſe that inhabit on
the South ſide the Equator, have their longeſt Day when the
Sun enters ♑. But to know how long the longeſt Day is in any
North or South Elevation, Raiſe the North or South Pole according to the Elevation of the Place, and bring ♋ for North
Elevation, or ♑ for South Elevation to the Meridian, and the
Index of the Hour Circle to 12. Then turn the Globe about
till ♋ for North Elevation, or ♑ for South Elevation, come to
the Weſt ſide the Horizon and the number of Hours and minutes pointed at on the Hour Circle, doubled, is the number of
Hours and minutes of the Longeſt Day.

The length of the Night to that Day is found by ſubſtracting
the length of the day from 24. for the remainder is the length of
the Night.

The

The ſhorteſt Day in that Latitude is the length of the ſhorteſt Night, found as before. And the longeſt Night is of the ſame length with the longeſt Day.

Example.

I would know the length of the longeſt Day at *London.* Therefore I Elevate the North Pole 51½ degrees, and bring ♋ to the Meridian, and the Index of the Hour Circle to 12. Then I turn ♋ to the Weſtern ſide the Horizon, and find the Index point at 8. hours 18. minutes, which being doubled makes 16. hours 36. minutes, for the length of the longeſt Day here at *London.*

PROB. XLIX.

To find how much the Pole is Raiſed, or Depreſſed, where the longeſt Day is an Hour longer or ſhorter then it is in your Habitation.

REctifie the Globe to the Latitude of your Place ; and make a prick at that point of the *Tropick* which is at the Meridian ; I mean at the *Tropick* of ♋, if your Habitation be on the North ſide the Equator ; or ♑, if your Habitation be on the South ſide the Equator : And if you would know where the longeſt Day is juſt an hour longer then it is at your Habitation, turn the Globe to the Weſtward till 7½ degrees of the Equator paſs through the Meridian, and make there another prick on the *Tropick* : Then turn about the Globe till the firſt prick come to the Horizon ; and move the Meridian through the notches of the Horizon till the ſecond prick on the *Tropick* come to the Horizon ; ſo ſhall the arch of the Meridian contained between the Elevation of your Place, and the Degree of the Meridian at the Horizon, be the number of Degrees that the Pole is Elevated higher then it is in your Latitude.

Example.

I would know in what Latitude the longeſt Day is an Hour longer then it is at *London.* Therefore I Rectifie the Globe to 51½ deg. and where the Meridian cuts the *Tropick* of ♋ I make
a prick

a prick ; then I note what degree of the Equator is at the Meridian, and from that degree on the Equator count 7 ½ degrees to the Eaſtwards, and bring thoſe 7 ½ degrees to the Meridian alſo; and again where the Meridian cuts the *Tropick* of ♋ I make another prick, ſo ſhall 7 ½ degrees of the *Tropick* be contained between thoſe two pricks. Then I turn the Globe about, till the firſt prick comes to the Horizon, and (with a Quill thruſt between the Meridian and the Ball) I faſten the Globe in this poſition: Afterwards I move the Meridian through the notches of the Horizon, till the ſecond prick riſes up to the Horizon, and then I find 56 ½ degrees of the Meridian cut by the Superficies of the Horizon : Therefore I ſay, In the Latitude of 56 ½ degrees, the longeſt Day is an Hour longer then it is here at *London.*

But if you would know in what Latitude the Dayes are an Hour ſhorter, you muſt make your ſecond prick 7 ½ degrees to the Weſtwards of the firſt, and after you have brought the firſt prick to the Horizon, you muſt depreſs the Pole till the ſecond prick deſcends to the Horizon : ſo ſhall the degree of the Meridian at the Horizon, ſhew in what Elevation of the Pole the Daies ſhall be an Hour ſhorter.

By this Probleme may be found the Alteration of Climates : for (as was ſaid in the Definition of (*Climates, Book* 1. *fol.* 28.) Climates alter according to the half-hourly increaſing of the Longeſt Day: therefore the Latitude of 56 ½ degrees having its Daies increaſed an whole Hour) is diſtant from the Latitude of *London* by the ſpace of two Climates.

PROB. L.

The Suns Place given, to find what alteration of Declination *he muſt have to make the Day an Hour longer, or ſhorter : And in what number of Daies it will be.*

REctifie the Globe to the Latitude of the Place, and bring the Suns place to the Eaſt ſide the Horizon, and note againſt what degree of the Horizon it is : then bring one of the Colures to interſect the Horizon in that degree of the Horizon, and at the point of Interſection make a prick in the Colure; and obſerve what degree of the Equator is then at the Meridian : Then turn the Globe Weſtward, if the Daies ſhorten; but Eaſtwards, if

O   they

they lengthen, till 7½ degrees of the Equator paſs through the Meridian, and where the Horizon interſects the ſame Colure, make another prick in the Colure : Afterwards bring the Colure to the Meridian, and count the number of degrees between the two pricks, for ſo many degrees muſt the Suns Dedination alter to lengthen or ſhorten the Day an Hour.

Example.

The Suns Place is ♉ 10. I would know how much he muſt alter his Declination before the Day is an Hour longer here at *London.* Therefore I rectifie the Globe to the Latitude of *London,* and bring ♉ 10. to the Eaſt ſide the Horizon, and find it againſt 24½ degrees from the Eaſt point : therefore I bring one of the Colures to this 24½ degrees, and cloſe by the edge of the Horizon I make a prick with black lead, in the Colure : then keeping the Globe in this poſition, I look what degree of the E-quator is then at the Meridian, and find 250¼, and becauſe the Daies lengthen, I turn the Globe Eaſtwards, till 7½ degrees from the foreſaid 250¼ paſs through the Meridian : then keeping the Globe in this poſition I make another prick in the Colure, and bringing this Colure to the Meridian, I find a little more then 5 degrees of the Meridian contained between the two pricks : therefore I ſay, when the Sun is in ♉ 10. degrees, he muſt alter his Declination a little more then 5 degrees, to make the Day an Hour longer.

Now to know in what number of Daies he ſhall alter this Declination, you muſt find the Declination of the two pricks on the Colure as you found the Suns Declination by Prob. 5. and the Arch of the Ecliptick that paſſes through the Meridian while the Globe is turned from the firſt pricks Declination to the ſecond pricks Declination, is the number of Ecliptical de-grees that the Sun is to paſs while he alters this Declination : and the degree of the Ecliptick then at the Meridian is (with reſpect had to the Quarter of the Year) the place the Sun ſhall have when its Declination ſhall be altered ſo much as to make the Day an Hour longer

Thus having the Suns firſt place given, and its ſecond place found: you may by finding thoſe two places on the Plain of the Horizon, alſo find the number of Daies comprehended between them, as you are taught by the fourth Probleme.

This

This Probleme thus wrought for different Times of the Year, will fhew the falacy of that Vulgar Rule which makes the Day to be lengthned or fhortned an Hour in every Fifteen Daies: when as the lengthning or fhortning of Daies keeps no fuch equality of proportion : for when the Sun is neer the Equinoctial points the Daies lengthen or fhorten very faft : but when he is neer the Tropical points, very flowly.

PROB. LI.

Of the Difference of Civil *and* Natural Daies, *commonly called the* Equation *of* Civil Daies. *And how it may be found by the* Globe.

THe Civil Day is that fpace of Time containing juft 24. Hours, reckoned from 12 a clock on one Day to 12 a clock the next Day ; in which fpace of Time the E-quinoctial makes upon the Poles of the World a Diurnal Revolution. The Natural Day is that fpace of Time wherein the Sun moveth from the Meridian of any Place to the fame Meridian again. Thefe Daies are at one time of the Year longer then at another ; and at all Times longer then the Civil Daies. There is but fmal difcrepancy between them, yet fome there is, made by a two-fold Caufe. For firft, The Suns Apparent motion is different from his true motion ; He being much flower in his *Apogeum* then he is in his *Perigeum* : For when the Sun is in his *Apogeum* he fcarce moves 58 minutes from Weft to Eaft in a Civil Day, but when he is in his *Perigeum* he moves above 61 minutes in a Civil Day : and therefore increafes his Right Afcenfion more in equal Time.

The fecond Caufe is the difference of Right Afcenfions anfwerable to equal parts of the Ecliptick: for about ♋ and ♑ the differences of Right Afcenfions are far greater then about ♈ and ♎ : for about ♈ and ♎ the Right Afcenfion of 10. degrees is but 9. degrees 11. minutes; but about ♋ and ♑ the Right Afcenfion of 10 degrees will be found to be 10. degrees 53. minutes, as by the Globe will appear.

But becaufe of the fmalnefs of the Globes graduation, you cannot actually diftinguifh to parts neer enough for the folution of this Probleme, if you fhould enquire the difference in length of two fingle Daies ; it will be requifite to take fome number of

O 2

Daies

Daies together ; Suppoſe 20. Therefore find by Prob. 3. the Places of the Sun for the beginning and ending of thoſe Daies you would compare ; and find the Right Aſcenſions anſwerable to each place in the Ecliptick ; and alſo the differences of Right Aſcenſions anſwerable to the Suns motion in each number of Daies : Then compare the differences of Right Aſcenſions toge-ther ; and by ſubſtracting the leſſer from the greater, you will have the number of degrees and minutes of the Equator that have paſſed through the Meridian more in one number of Daies then in the other number of Daies : which degrees of the Equa-tor converted into Time, is the number of minutes that the one number of Daies is longer then the other number of Daies.

Example.

I would know what difference of Time there is in the length of the firſt 20. Daies of *December,* and the firſt 20, Daies of *March.* I find by Prob. 3. the Suns place *December* 1, is ♐ 19. 45. at the end of 20 Daies. *viz.* on the 21 Day his place is ♑ 10. 11. The Suns place *March* 1. is ♓ 21. 16. at the 20. Daies end, *viz. March* 21, his place is ♈ 11. 3.

I find by Prob. 26. the
Right Aſcenſion anſwerable
to

$$\left\{\begin{array}{ll}♐ & 19. \quad 45 \\ ♑ & 10. \quad 11 \\ ♓ & 21. \quad 16 \\ ♈ & 11. \quad 3\end{array}\right\} \text{ is } \left\{\begin{array}{ll}258. & 10. \\ 280. & 25. \\ 352. & 00. \\ 9. & 40.\end{array}\right.$$

and the difference of Right Aſcenſions contained between the firſt Day in each Moneth, and the 21 of the ſame Moneth, by ſubſtracting the leſſer from the greater is for

$$\left\{\begin{array}{l}258. \quad 10. \\ 280. \quad 25. \\ \overline{22. \quad 15}\end{array}\right. \quad \text{And for} \quad \left\{\begin{array}{l}352. \, 00. \\ 9. \, 40. \\ \overline{17.40.}\end{array}\right.$$

But note, becauſe the Vernal Colure, where the degrees of Right Aſcenſion begin and end their account, is intercepted in the Arch of the Suns motion from the firſt to the 21. of *March,* therefore inſtead of ſubſtracting the leſſer number of degrees of Right Aſcenſion from the greater, *viz.* 9. 40 from 35. 2. I do for finding the difference of the Right Aſcenſional arch of the Suns motion in thoſe 20 Daies, ſuſtract the foreſaid 352 degrees from 360, and the remains is 8. which is the difference of Right Aſ-cenſion from ♓ 21, 16. to the Equinoctial Colure: to which 8 adding

adding 9 degrees 40 minutes, the Right Afcenfion from the E-quinoctial Colure to ♈ 11. 3. it makes 17 degrees 40. minutes for the difference of Right Afcenfions between ♓ 21. 16. and ♈ 11. 3 Then I find the difference of this Difference of Right Afcenfion, by fubftracting the lefs from the greater, *viz.* 17. 40. from 22. 15. and the remains is 4. degrees 35. minutes, for the number of degrees and minutes of the Equator that pafs through the Meridian in the firft 20 Daies in the Moneth of *December* more then in the firft 20 Daies of the Moneth of *March*: which 4. degrees 35. minutes converted into Time, gives 19. minutes, that is, a quarter of an Hour and 4 minutes that the firft 20 Daies of *December* aforefaid, are longer then the firft 20 Daies of *March*.

PROB. LII.

How to find the Hour of the Night, when the Moon *fhines on a* Sun Dyal, *by help of the* Globe.

REctifie the Globe, and find by Prob. 54. or an *Ephemeris*, the Moons place at Noon : Bring it to the Meridian, and the Index of the Hour Circle to 12. and turn about the Globe till the Index of the Hour Circle points to the fame Hour the fhade of the Moon falls on, on the Sun Dyal. Then by Prob. 3. find the Suns place at Noon, and fee how many degrees of Right Afcenfion are contained between the Suns place and the degree of the Equator at the Meridian, when the Index of the Hour Circle is brought to the Hour the Moon fhines on in the Sun Dyal ; for thofe number of degrees converted into Time, fhall be the Time from Noon, or the Hour of the Night. Only note, Refpect muft be had to the motion of the Moon from Weft to Eaft, for fo fwift is her mean motion, that it is accounted to be above 12. degrees in 24. Hours ; that is 6 degrees in 12 Hours, 3 degrees in 6 Hours, &c. and this alfo converted into Time, as aforefaid, you muft add proportionably to the Time found from Noon; and the fum fhall give you the true Hour of the Night.

Example.

Here at *London*, I defired to know the Hour of the Night *Ja-nuary* 6. this prefent Year 1658. The Moons place found by

O 3   an an *Ephemeris,* or for want of an *Ephemeris,* by Prob. 54. is in ♓ 21. degree 22. minutes; Therefore I rectified the Globe to *Londons* Latitude, and brought ♓ 21. 22. minutes to the Meridian, and the Index of the Hour Circle to 12. then by Prob. 3. I found the Suns place in ♑ 26. degrees 46. minutes, and by Prob. 26. I found his Right Aſcenſion to be 300 degrees; Then I turned about the Globe, till the Index of the Hour Circle pointed at 10 Hours, and at the degree of the Equator at the Meridian I made a prick ; then I counted the number of degrees of the Equater contained between the foreſaid 300 deg. and this prick and found them 111 $\frac{1}{4}$ degrees which converted into Time, by allowing 15 degrees for an Hour, gives 7 hours, 25 minutes, Time from Noon : which if the Moons motion were not to be conſidered, ſhould be the immediate Hour of the Night: But by the Rule aforeſaid, the Moons motion from Weſt to Eaſt, in 7 hours 25 minutes is 3 degrees 42 minutes, and this 3 degrees 42 minutes being converted into Time, is 14 minutes more, which being added to 7 hours 25 minutes : make 7 hours 39 minutes, for the true Hour of the Night.

PROB. LIII.

To find the Dominical Letter, *the* Prime, Epact, Eaſter Day, *and the reſt of the* Moveable Feaſts, *for ever.*

THough theſe Problemes cannot be performed by the Globe, becauſe of the ſeveral changes, and irregular accounts that their Rules are framed upon, yet becauſe they are of frequent and Vulgar uſe, and for that the ſolution of many other Queſtions will have dependency on the knowledge theſe ; Therefore I have thought fit here to inſerte this Table of M*r* *Palmers,* by which you may find them All.

I ſhall not inſiſt upon the Reaſons of the ſeveral changes of Letters, and Numbers, Himſelf having already very learnedly handled that ſubject, in his Book of the *Catholick Planiſphear, Book* 1. *Chapter* 11. (to which I refer you) Neither ſhall I need to give you any other Inſtructions for finding what is here propoſed, then what himſelf hath given in his fourth *Book, Chapter 66,* and part of *67.* Therefore take it as he there delivers it.

An Exam-

An Example, fhall ferve here inftead of a Rule. For the Year 1657. I would know all thefe: wherefore I feek the Year 1657. in the Table of the Suns *Cycle,* and over againft it, I find 14. for the Year of the *Cycle* of the Sun, and D for the *Dominical Letter.* And note here, that every *Leap-year* hath 2 *Dominical Letters* (as 1660, hath A G) and the firft (*viz.* A) ferveth that Year till *February* 25, and the fecond (G) for the reft of the Year. And note that thefe Letters go alwayes backwards when you count forwards (as B A, then G F, &c. not F G, and then A B) as you may fee by the Table.

Then in the Table of the *Cycle* of the Moon, I have for the Year 1657. the *Prime* 5. the *Epact* 25. Thofe had, I go to the Table for *Eafter,* and feek there in the firft rank the *Prime* 5. and

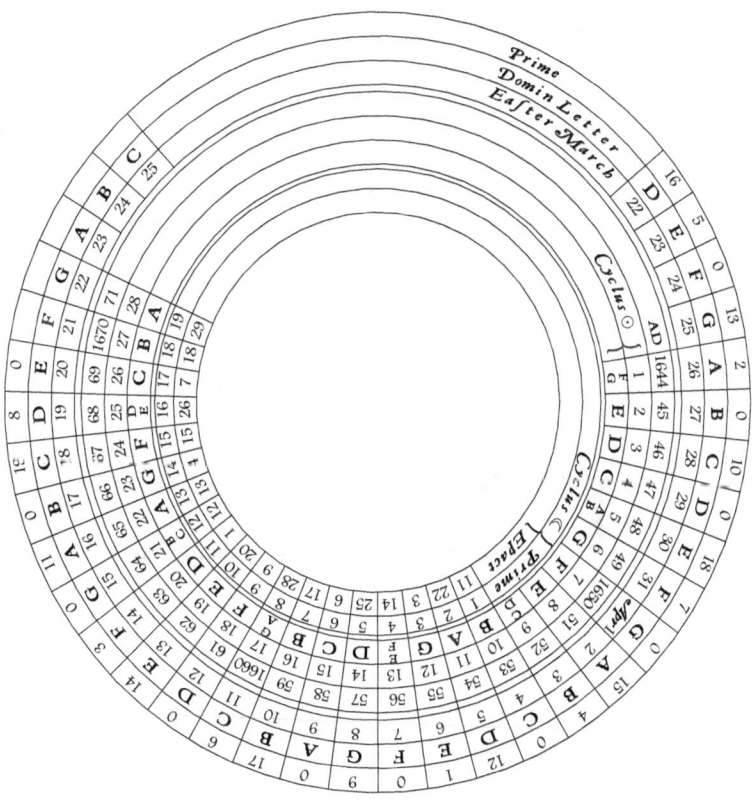

under

under it in the middle rank ſtands E; that is not my *Dominical Letter;* therefore I ſeek not backward, but alwayes forward in the middle rank, till I come to my *Dominical Letter* D. and under it I find in the third rank *March* 29. upon which *Eaſter* day falls this Year 1657. The reſt of the moveable *Feaſts* may be had by their diſtances from *Eaſter,* which are always the ſame. One-ly for *Advent Sunday,* remember that the next Sunday after *November* 26 is *Advent Sunday.* Read Book 1. 11. and that will ſufficiently inſtruct you with this Example.

To find the Age of the Moon.

Remember firſt that the *Epact* begins with *March,* which muſt be here accounted the firſt Moneth : Then if you add to the *Epact* the number of the Moneth current, and the number of the day of the Moneth current, the ſum or the exceſs above 30. is the Moons age.

Example. *January* 20. 1656. According to the accompt of the Church of *England,* (who begin the Year with *March* 25. which was the Equinoctial day about Chriſt time) the *Epact* is 14. *January* is the 11 *th* Moneth, and the 20 *th* day is propoſed; now add 14. 11. and 20. together, they make 45. out of which I take 30. and there remains 15. the Moons age.

PROB. LIV.

The Age of the Moon *given, to find her place in the* Eclip-tick *according to her mean motion.*

THis Probleme may be performed exact enough for Com-mon uſes by the Globe, but in regard it only ſhews the Moons place in the *Ecliptick* according to her mean motion, it will often fail you ſome few degrees of her true Place. The work is thus,

Firſt ſet figures to every twelfth degree of the Equinoctial, ac-counted from the Equinoctial Colure, marking them with 1, 2, 3, 4. &c. to 30 which will end where you began *viz.* at the E-quinoctial Colure again : ſo ſhall the Equinoctial be divided into 30 equal parts, repreſenting the 30 Dayes of the Moons Age Theſe figures (to diſtinguiſh them from the degrees of the Equator) were beſt be writ with Red Ink.   When

When you would enquire the Moons Place, Elevate the North Pole 90 degrees, that is, in the *Zenith,* so shall the Equator ly in the Horizon : Then bring the Equinoctial Colure against the Day of the Moneth in the Horizon, so shall the Moons Age written in Red figures, stand against the Signe and degree in the Horizon that the Moon is in at that Time.

Example.

September 28. 1658. I would know the Moons place in the *Ecliptick,* she being then 12 Daies old. Therefore I Elevate the North Pole 90 degrees above the Horizon, and turn the Globe about till the Equinoctial Colure come to *September* 28. in the Circle of Daies on the Horizon ; then looking against what Signe and degree of the *Ecliptick* Circle in the Horizon the 12 ᵗʰ division in Red figures stands, I find ♓ 9. which is the Signe and degree the Moon is in, according to her mean Motion.

This Probleme may be applyed to many Uses: for, having the Moons Place you may find the Time of her Rising, Southing, Setting, and Shining &c. by working with her, as you were taught to work with the Sun, in several fore-going Problemes, proper to each purpose.

PROB. LV.

Having the Longitude *and* Latitude, *or* Right Ascension *and* Declination *of any* Planet, *or* Comet, *to place it on the* Globe, *to correspond with its place in* Heaven.

Planets and Comets cannot be placed on the Globe so as their places will long retain correspondence with their places in Heaven ; Because as was said (*Chap.* 44. they have a continual motion from West to East upon the Poles of the *Ecliptick:* yet never-the-less you may by having their Longitude and Latitude, or Right Ascension and Declination, for any set Time, place a Mark for them on the Globe, either with Ink if your Globe be Varnisht, for then you may with a wet finger wipe it off again ; or with Black-lead, if it be not Varnisht, and then you may rub it out again with a little White

P   Bread :

Bread : which Mark for that Time, will as effectually ferve you to work by, as any of the Fixed Stars placed on the Globe will do.

Therefore if the Longitude and Latitude of any Planet, or Comet, be given ; Do thus, Elevate the North Pole, if the Latitude given be North ; but if the Latitude given be South, Elevate the South Pole *66* ½ degrees; and place the Pole of the *Ecliptick* in the *Zenith,* and over it fcrew the Quadrant of Altitude : fo fhall the *Ecliptick* ly in the Horizon ; and the Quadrant of Altitude being turned about the Horizon fhall pafs through all the Degrees of Longitude : Then find the point of given Longitude in the *Ecliptick,* and bring it to the Quadrant of Altitude, and hold it there : Then count upwards on the Quadrant of Altitude the number of degrees and minutes of given Latitude, and at the point where the number ends, clofe to the Quadrant of Altitude, make a fmal Prick, and that Prick fhall reprefent the Planet or Comet you were to place on the Globe.

If it be the Right Afcenfion and Declination of a Planet or Comet that is given ; you muft find the degree and minute of Right Afcenfion on the Equinoctial, and bring it to the Meridian, and keep the Globe there fteddy ; then find the degree and minute of Declination on the Meridian, and under that degree and minute on the Globe make a Prick, and that Prick fhall reprefent the Planet, or Comet, as aforefaid,

If it be ♄ or ♃ that this Prick is to reprefent, it may ftand on the Globe fometimes a Week or a Fortnight, without much difference from the Planets place in Heaven. But if the Prick were to reprefent the other Planets, you muft (in regard of their fwift motion) alter it very often, efpecially for the Moon ; for fo fwift is her motion, that in every two Hours fhe alters about a degree in Longitude.

Having thus placed this Mark on the Globe, you may find out the Time of its feveral Pofitions, and Afpects, if you work by it as you are directed to work by the Sun, in the feveral refpective Problemes throughout this Book.

The End of the Second Book.

The Third BOOK,

Being the Practical Use of the

GLOBES.

Applyed to the Solution of Problemes

In the Art of

NAVIGATION.

PRÆFACE.

Ecaufe the Art of Navigation confifts afwell in the knowledge of Aftronomical *and* Geographical Problemes, *as in Problemes meerly Nautical ;* Therefore I muft defire the Artift to feek in the laft Book fuch Problemes as are only Aftronomical or Geographi-cal. For my Defigne is here to collect fuch Problemes as are only ufed in the Art of Navigation, fome few particulars ex-cepted, as for finding Latitude, Longitude, Courfe; Di-ftance, &c. Which though they are handled in that Book, yet for their exceeding Vtility in the Art of Navigation, and for that what there is given, cannot always be had to work by ; therefore in this Book I have mentioned divers other Obfervations, which being made or had, you may by the Rules proper for each Obfervation find what fhall be pro-pofed. P 2 PROB.

PROB. I.

The Suns Amplitude *and* Difference *of* Afcenfion *given, to find the* Heigth *of the* Pole, *and* Declination *of the* Sun.

Elevate the Pole fo many degrees as the Difference of the Suns Afcenfion is, and fcrew the Quadrant of Altitude to the *Zenith*, and bring the firft point of ♈ to the Meridian, then number on the Quadrant of Altitude upwards the complement to 90. of the Suns Amplitude, and move the Quadrant of Altitude till that number of degrees cuts the Equator ; So fhall the Quadrant cut in the Horizon the degree of the Pole Elevation; and in the Equator the degree of the Suns Declination.

Example.

The difference of Afcenfion is 27. degrees 7. minutes. Therefore I Elevate the Pole 27. degrees 7. minutes above the Horizon, and fcrew the Quadrant of Altitude to 27. degrees 7. minutes, which is in the *Zenith* : then I bring the firft point of ♈ to the Meridian, and number on the Quadrant of Altitude upwards 56. degrees 40. minutes, the Complement of the Suns Amplitude, and bring that degree to the Equator ; then I fee in what degree of the Horizon the Quadrant cuts the Horizon, and find $51\frac{1}{2}$, which is the Elevation of the Pole: then looking in what degree of the Equator the Quadrant of Altitude cuts the Equator, I find 20 degrees, 5 min. which is the Declination of the Sun at the fame Time.

PROB. II.

The Suns Declination *and* Amplitude *given, to find the* Poles Elevation.

Elevate the Pole fo many degrees as the Complement of the Suns Amplitude is ; and fcrew the Quadrant of Altitude

in

in the *Zenith*, and bring the firſt point of ♈ to the Meridian : Then count on the Quadrant of Altitude to the Degree of the Suns Declination, and bring that degree to the Equinoctial; and the degree of the Equinoctial cut by that degree of the Quadrant of Altitude, is the degree of the Poles Elevation.

Example.

The Suns Amplitude is 33. degrees 20. minutes, his Declination is 20 degrees 5 minutes, his Complement of Amplitude to 90. is 56 degrees 7 minutes. Therefore I Elevate the Pole 56. degrees 7 minutes above the Horizon, and ſcrew the Quadrant of Altitude to 56 degrees 7 minutes which is in the *Zenith* : Then I bring the firſt point of ♈ to the Meridian, and number on the Quadrant of Altitude upwards 20. deg. 5 min. for the Suns Declination, this 20*th* degree 5 minutes, I bring to the Equinoctial, and find it cut there 51 ½. degrees, for the Heigth of the Pole.

PROB. III.

The Suns Declination *and* Hour *at* Eaſt *given, to find the* Heigth *of the* Pole.

E Levate the Pole ſo many degrees as the Suns Declination is, and ſcrew the Quadrant of Altitude in the *Zenith* : Then convert the Hours or minutes paſt 6. given into degrees ; by allowing 15 degrees for every Hour of Time, and for every minute of Time 15 minutes of a Degree ; and number thoſe degrees or minutes in the Horizon from the Eaſt Southwards; ſo ſhall the Degree of the Quadrant of Altitude cut by the Equator be the Complement of the heigth of the Pole.

Example.

The Suns Declination is 20 deg. 5 min. Therefore I Elevate the Pole 20 degrees 5 minutes, and alſo ſcrew the Quadrant of Altitude to 20 degrees 5 minutes which is in the *Zenith*: the Hour the Sun comes to be at Eaſt is 8 a clock 53 minutes, that is, 1 Hour 7 minutes after 6. Therefore I convert 1 Hour 7 minutes

P 3   nutes

nutes into Degrees, as before, and it gives 16 degrees 50 minutes; which number of degrees and minutes I count from the East point Southwards, and thither I bring the Quadrant of Altitude : Then I look in what degree of the Quadrant of Altitude, the Equator cuts, and find 38 $\frac{1}{2}$, which is the Complement of the Poles Heigth, *viz.* 51 $\frac{1}{2}$ degrees for the Heigth of the Pole.

In this Probleme the Declination of the Sun and Elevation of the Pole bears the same Denomination of either North or South, for when the Declination and the Elevation are different the Sun cannot come to the East point.

PROB. IIII.

The Declination *of the* Sun *and his* Altitude *at* East *given, to find the* Heigth *of the* Pole.

ELevate the Pole to the Complement of the Suns Altitude, and screw the Quadrant of Altitude to the *Zenith* : Then bring the Equinoctial point ♈ to the Meridian, and number on the Quadrant of Altitude the degrees of the Suns Declination, and bring that degree to the Equinoctial, and note the degree it cuts ; for its Complement to 90 is the Heigth of the Pole.

Example.

May 10. The Suns Declination is 20 degrees 5 minutes; His Altitude at East is 25 degrees 55 minutes here at *London:* I enquire the Heigth of the Pole. Therefore I substract 20. 5 min. from 90 the remains is 69 deg. 55 min. for its Complement; wherefore I bring 69 deg. 55 min. of the Meridian to the Horizon; and to 69 deg. 55 min. which is in the *Zenith,* I screw the Quadrant of Altitude then I bring ♈ to the Meridian, and count on the Quadrant of Altitude upwards 20 deg. 5 min. and move it about the Equinoctial till those 20 deg. 5 min. touch the Equinoctial, which I find to be in 38 $\frac{1}{2}$ degrees, Therefore I substract those 38 $\frac{1}{2}$ from 90, and the remains is 51 $\frac{1}{2}$ degrees. Therefore I say the Pole here at *London* is Elevated 51 $\frac{1}{2}$ degrees.

The Declination and the Elevation is alwaies the same, either North or South, for when they alter their Denominations the Sun at East can have no Altitude, neither can it indeed reach the

East

Eaſt point : and therefore in this example, becauſe the Declination of the Sun is North, it is the North Pole that is Elevated here at *London.*

To perform the ſame otherwiſe, with a pair of Compaſſes.

Take off with your Compaſſes from the Equator or Quadrant of Altitude the number of degrees of Altitude obſerved, and place one foot at the beginning of ♈ on the inner edge of the Horizon, and extend the other directly upwards towards the *Zenith* : Then move the Brazen Meridian through the notches of the Horizon till the other point of your Compaſſes (reſpecting the *Zenith*) reach the Parallel of the Suns Declination : So ſhall the number of degrees on the Meridian be the number of degrees that the Pole is Elevated above the Horizon ; and is either North or South according as the Suns Declination is : as before.

This may yet otherwiſe be performed with the Quadrant of Altitude, by taking the Nut off the Meridian, and laying the edge of its Index (ſpecified in *Chap.* 1. *Sect.* 6. of the firſt *Book*) exactly on the Eaſt line of the Horizon: for when that lies ſtraight between the point of Eaſt on the outer Verge of the Horizon, and the beginning of ♈ in the inner Verge of the Horizon, then ſhall the upper end of the Quadrant of Altitude point directly to the *Zenith:* and if then you turn the Meridian through the notches of the Horizon till the Suns Altitude on the Quadrant of Altitude cut the Parallel of Declination, you will have on the Meridian the heigth of the Pole : as before.

PROB. V.

By the Suns Declination *and* Azimuth *at 6. of the Clock given, to find the Heigth of the* Pole, *and* Almicantar *at 6.*

ELevate the Pole ſo many degrees as the Suns *Azimuth* is at 6. and ſcrew the Quadrant of Altitude in the *Zenith,* and bring the firſt point of ♈ to the Meridian : Then number on the Quadrant of Altitude upwards the Complement of the Suns Declination, and bring that degree to the Equator: So ſhall the

degree

degree of the Horizon cut by the Quadrant of Altitude be the Complement of the Poles Elevation; and the degree of the Equator cut by the Quadrant of Altitude ſhall be the *Almicantar* of the Sun at *6.* of the clock.

Example.

The Suns *Azimuth* at *6* is 12 ¾ degrees : Therefore I Elevate the Pole 12 ¾ , and ſcrew the Quadrant of Altitude to 12 ¾ degrees which is in the *Zenith* : Then I bring the firſt point of ♈ to the Meridian ; The Suns Declination is 20 degrees 5 minutes. Therefore I number on the Quadrant of Altitude *69* deg. 55 min. which is the Complement of 20 deg. 5 min. to 90. this *69* deg. 55 min. on the Quadrant of Altitude I bring to cut the Equator, and find when *69* deg. 55 min. cuts the Equator, that the Quadrant of Altitude cuts the Horizon, in 38 ½ deg. which is the Complement of the Poles Elevation: and at the ſame time the Quadrant of Altitude alſo cuts the Equator in 15 ½ degrees which is the *Almicantar* or Altitude of the Sun at *6.* a clock.

PROB. VI.

By the Hour *of the* Night *and a* known Star *Obſerved* Riſing *or* Setting, *to find the* Heigth *of the* Pole.

REctifie the Hour Index, by Prob. 2. of the former Book; and turn the Globe Weſtwards till the Hour Index points at the Hour of the Night; faſten the Globe there, and turn the Meridian through the notches of the Horizon till the known Star come to the Eaſt ſide the Horizon, if the Star be Riſing, or the Weſt if it be Setting; ſo ſhall the degrees of the Poles Elevation be cut by the Horizon under the Elevated Pole; and is North or South according as the Elevated Pole of the Globe is.

PROB. VII.

Two Places *given in the ſame* Latitude, *to find the* Difference *of* Longitude.

BRing the firſt Place to the Meridian, and note the number of degrees of the Equinoctial that comes to the Meridan
with

with it; then Bring the other place to the Meridian and note the
number of degrees of the Equator that comes to the Meridian
with it : and by fubftracting the leffer number from the greater
you have the difference of Longitude. This needs no Example.

PROB. VIII.

Two Places given in the fame Longitude, *to find the Difference of* Latitude.

B Ring the Places to the Meridian, and the degrees of the
Meridian over the two Places is the Latitudes of them
both, and by fubftracting the leffer number of degrees
from the greater you will have the difference of Latitude.

PROB. IX.

Courfe *and* Diftance *between two Places given, to find
their Difference in* Longitude *and* Latitude.

S Eek the Rhumb you have failed upon, as in Prob. 34. of the
laft Book, and upon that Rhumb make a mark for the
Place you departed from; then with your Compaffes take
off from the Equinoctial the number of Leagues you have
failed upon that Rhumb, by allowing a degree for every 20.
Leagues and place one foot of your Compaffes upon that mark,
and where the other foot falls on that Rhumb make a fecond
mark; then by bringing the firft mark to the Meridian, you will
fee on the Meridian the Latitude of that mark, and in the Equator the Longitude as in Prob. 2. of the laft Book: and by bringing
the fecond mark alfo to the Meridian, you will as before, find the
Longitude and Latitude of the fecond mark alfo. Then by fubftracting the leffer Latitude from the greater Latitude, and the
leffer Longitude from the greater Longitude, you will have the
difference remaining, both of Longitude and Latitude you are a-
rived into.

Q   PROB.

PROB. X.

To find how many Miles *are contained in a* Degree *of any* Parallel.

EVery Degree of the Equinoctial contains 20. Engliſh Leagues and every League 3. Engliſh Miles : But in e-very Parallel to the Equinoctial, the Degrees diminiſh more and more even to the Pole, where they end in a point. Therefore a Degree in any Parallel cannot contain ſo many Miles as a De-gree in the Equinoctial. Now that you may know how many Miles are contained in a Degree of any Parallel to the Equi-noctial. Do thus, Meaſure with your Compaſſes the width of any number of Degrees in any given Parallel; ſuppoſe (for Exam-ples ſake) 10. Degrees in the Parallel of 51 ½ ; Examine in the E-quator, how many Degrees of the Equator they will make, and you will find 6 ⅕. Therefore 1. Degree in the Equator making 60 Miles 6. Degrees makes 360, to which add for the ⅕ part 12 Miles, makes 372 Miles, to be the Meaſure of 10 Degrees in the Parallel of 51 ½. So that by dividing 372. by 10. you have 37 Miles for the length of a Degree, from Eaſt to Weſt in the Parallel of 51 ½ Degrees.

PROB. XI.

The Rhumb *you have ſailed upon, and the* Latitudes *you departed from, and are arived to, given, to find the* Difference *of* Longitude, *and the number of Leagues you have Sailed.*

FIrſt ſeek the Rhumb you have ſailed on, and paſs it through the Meridian till it cuts in the Meridian the La-titude you departed from; and keeping the Globe there ſtedey make a mark cloſe by the Meridian, under that Latitude and in that Rhumb on the Globe, and note in the E-quinoctial the degree of Longitude at the Meridian: then paſs that Rhumb through the Meridian again, till it cuts in the Meridian the Latitude you are arived to; and in that Rhumb and Latitude make on the Globe another mark, and examine in the Equinoctial

the

the Longitude of the second mark; for the difference between the first and second mark is the difference of Longitude. Then open your Compasses to one Degree of the Equinoctial, and by measuring along in the Rhumb count how many times that Distance is contained between the two points in that Rhumb: for so many times 20. Leagues is the Distance you have sailed.

Example.

I sail upon the North West Rhumb from the Latitude of 10. degrees, into the Latitude of 30. degrees 40. minutes. Therefore I find the North West Rhumb, and turn the Globe through the Meridian till this Rhumb cut the Meridian in the first Latitude, *viz.* in 10. degrees and directly under 10. degrees upon the Rhumb I make a prick, and also find 10 degrees 3 minutes, of the Equator at the Meridian, for the Longitude of the First Place. Then I turn the Globe again through the Meridian, till the same Rhumb cut the Meridian in the second Latitude, *viz* in 30 degrees 40 minutes, and directly under those 30 degrees 40 minutes upon the same Rhumb, I make another prick, which represents the Place I am arrived to : I examine the Longitude of this prick, as before, and find it 32 degrees 10 minutes. Therefore I substract the first Longitude, *viz.* 10 degrees 3 minutes from the second Longitude, *viz.* 32 degrees 10 minutes, and there remains 22 degrees 7 minutes, for the Difference of Longitude.

Then for examining the Distance I open my Compasses to 1. degree on the Equinoctial and measure upon the Rhumb how oft that Distance is contained between the two pricks, and find $29\frac{1}{4}$, that is, 29 degrees 15 minutes, which multiplyed by 20. gives 585. for the number of Leagues sailed upon that Rumb.

The reason why I open the Compasses no wider then to 1 degree, is because the Rhumbs being Circular or crooked lines the distance on them may be measured more exactly by often counting that 1 degree in them then if the Compasses had bin opened to many degrees. Thus if the Compasses had been opened wide enough to reach between the two pricks aforesaid, I should not have had above 583 Leagues for the distance between the two Places : neither is there indeed more great Circle distance between them ; But I sailed upon a Rhumb, that is, I followd

the

the Courſe of a Circular winding line, and ſo fetcht a Compaſs about to come to theſe two pricks; and therefore I have in truth ſailed 585. Leagues. For the ſegment of a Rhumb between two Places is alwaies greater then a ſtraight line drawn betwixt them; yea ſometimes by half or more in Places neer either Pole.

Note, If you be not very curious in opening your Compaſſes to this ſmal diſtance, you may in oft turning them about upon the Rhumb commit error in your meaſuring: therefore when you have taken the Diſtance of one degree, try if you neither gain or looſe any thing in meaſuring 10, or 20. degrees of the Equi-noctial by them: for then your Compaſſes are opened to a width exact enough for your purpoſe.

PROB. XII.

The Longitudes *and* Latitudes *of two* Places given, *to* find Courſe, *and* Great Circle *diſtance between them.*

FInd on the Globe the Longitudes and Latitudes given, and make pricks to either Longitude and Latitude : If any Rhumb paſs from one place to the other, that is (without more adoe) the Rhumb ſought. But if no Rhumb paſs through ; Take the Rhumb that runs moſt Parallel to the two pricks: for that ſhall be the Rhumb or the neereſt Rhumb that theſe two pricks Bear on. An Example of this, ſee in Prob. 34. of the Laſt Book : And the Great Circle Diſtance between theſe two pricks, you may find as by Prob. 33. of the ſame Book.

PROB. XIII.

The Latitude *you departed from, and the* Latitude *you are arrived to, and the number of* Leagues *you have ſailed given, to find the* Rhumb *you have ſailed on, and diffe-rence of* Longitude.

MAke a prick on the Globe in the Latitude you departed from: then open your Compaſſes to the number of Leagues you have ſailed, by taking for every 20. Leagues 1. degree of the Equator, half a degree for

10.

10 Leagues, a quarter of a Degree for 5 Leagues, and fo proportionably for any other number of Leagues. Place one foot of your
Compaffes in the prick made for the Latitude you departed from,
and extend the other towards the Latitude you are arived to, and
difcribe an occult Arch ; Turn the Globe till this occult Arch
come to the Latitude on the Meridian, and where the Latitude
cuts this occult arch make another prick to reprefent the Latitude
you are arived to; fo fhall the Rhumb paffing through thofe two
pricks (or that is moft Parallel to thofe two pricks) be as in the
laft Prob. the Courfe or the Rhumb thofe two pricks Bears on.

The difference of Longitude you may find as by Prob. 11.

PROB. XIV.

To find by the Globe the Variation of the Needle; *commonly called the* Variation of the Compafs.

Bferve by a Compafs whofe wyer is placed juft under
the *Flower deluce,* what point of the Compafs the Sun
Rifes or Sets on, Morning, or Evening : Then examine by Prob: 10. of the fecond Book, what degree of
the Horizon the Sun Rifes or Sets on by the Globe alfo; and if
the Rifing or Setting be the fame, both on the Globe and Compafs, there is no Variation in your Place, But if there be difference
between the Rifing or Setting by the Compafs and the Globe,
then is there Variation in your Place.

If the point the Sun Rifes upon in the Compafs be neerer the
North point, then the point the Sun Rifes upon by the Globe, the
Variation is Weftwards.

If the point the Sun Sets upon in the Compafs be neerer the
North then the point it Sets upon by the Globe, the Variation
is Eaftwards.

If the point the Sun Sets upon in the Compafs be further from
the North point, then the point the Sun Sets upon by the Globe,
the Variation is Weftwards.

If the point the Sun Rifes upon in the Compafs be further
from the North point then the point the Sun Rifes upon by the
Globe, the Variation is Eaftwards. And fo many degrees as there
is between the point of Rifing or Setting found by the Compafs,
and the point of true Rifing or Setting found by the Globe, fo

Q3   many

many degrees is the Variation from the North towards the Eaſt, or Weſt point.

Otherwiſe, when the Sun hath Altitude.

Having the Altitude of the Sun ; find by Prob. 22. of the ſecond Book, its *Azimuth* : Then examine by a Compaſs whether the true *Azimuth* found by the Globe, agree with the *Azimuth* found by a Nautical Compaſs : If they agree there is no Variation : But if the *Azimuth* of the Compaſs before Noon be neerer the North then the true *Azimuth* found by the Globe, the Variation is Weſtwards.

If the *Azimuth* by the Compaſs Afternoon be neerer the North, the Variation is Eaſtwards.

If the *Azimuth* by the Compaſs Afternoon be further from the North, the Variation is Weſtwards,

If the *Azimuth* by the Compaſs before Noon be further from the North, the Variation is Eaſtwards.

And this Variation ſhall be as aforeſaid ſo many degrees as there is between the *Azimuth* Obſerved by the Compaſs, and the true *Azimuth*, Obſerved by the Globe.

PROB. XV.

To keep a Journal *by the* Globe.

BY ſome of theſe foregoing Problemes you may Dayly (when Obſervations can be made) find both the Longitude and Latitude on the Globe of the Places you are arived to, and alſo the Way the Ship hath made, and make pricks on the Globe in their proper Places for every Daies Journey, ſo truly and ſo naturally that if you kept your reckoning aright you may be ſure you cannot miſs any thing of the truth it ſelf ; and that with leſs trouble and greater advantage, then keeping a Book of every Daies Reckoning.

PROB.

PROB. XVI.

To Steer *in the* Night *by the* Stars.

Rectifie the Globe and Hour Index as by Prob. 2. of the laſt Book, and turn about the Globe till the Index of the Hour Circle points to the Hour of the Day or Night: Then turn the Globe till the Difference of Longitude between the Place you depart from, and the Place you ſail to paſs through the Meridian and if any Star in the Latitude of the Place you ſail To come to the Meridian, or neer the Meridian with the degree of the difference of Longitude, that Star is at that time in or neer the *Zenith* of that Place you ſail to: and by finding the ſame Star in Heaven, as by Prob. 44. of the laſt Book you may direct your ſhip towards that Star, and ſail as confidently (ſaies Mr *Blagrave*) as if *Mercurie* were your Guide. But becauſe this Star moves from the *Zenith* of this Place you muſt often examine what Star is come to the *Zenith*, and ſo often charge the Star you Steer by, as the length of your Voyage may require.

PROB. XVII.

How to platt on the Globe *a* New Land, *never before* Diſcovered.

Theſe two following Problemes are 2. Chapters of Mr *Wrights*, delivered by him as follows.

It may ſometimes fall out in new Diſcoveries, or when your Ship by means of a Tempeſt is driven out of her right Courſe, that you ſhall come to the ſight of ſome Iſle, Shoald, or new Land, whereof the Mariner is utterly ignorant. And to make ſome relation of the ſame, or to go unto it ſome other time, if you deſire to ſet it down on your Globe in the true place, you may do it after this manner : So ſoon as you have ſight thereof, mark it well firſt with your Compaſs, obſerving diligently upon which Point thereof it lieth. And ſecondly, you muſt there take the heigth of the Sun, or of the Pole-ſtar, as you were taught Prob. 13. of the ſecond Book, that you may know

in

in what Point your Ship is, and that point you muſt call the Firſt
Point; which being ſo done, your Ship may ſail on her Courſe
all that day, till the day following, without loſing her Way: and
the next day mark the Land again, and ſee upon what Point it lieth; and then take your heigth, and with it caſt your Point of
Traverſe once again ; and that you may call your ſecond Point.
Then take a pair of Compaſſes, and placing one foot upon the
Firſt Point, and the other upon the Rhumb towards which the
Land did Bear, when you Caſt your Firſt Point : ſet alſo one foot
of another pair of Compaſſes in the ſecond Point, and the other
foot upon the Rhumb upon which the Land lay when you caſt
your ſecond Point; and theſe two Compaſſes thus opened, you
muſt move by their Rhumbs, till thoſe two feet of both Compaſſes do meet together, which were moved from the foreſaid
two Points : and where they do ſo meet together, there may
you ſay is the Land which you Diſcovered; which Land you
may point out with the In lets and Out-lets, or Capes and other
Signes, which you ſaw thereupon. And by the graduation you
may ſee the Latitude thereof; that thereby you may find it, if at
any time after you go to ſeek for it.

PROB. XVIII.

Seeing two known Points *or* Capes *of Land, as you ſail a-
long, how to know the diſtance of your Ship from them.*

Itch one foot of one pair of Compaſſes upon one of the
two foreſaid Capes, and the other foot upon the Rhumb
which in this Compaſs pointeth towards that Cape. And
in like manner ſhall you do with another pair of Compaſſes, placing one foot thereof upon the other known Cape, and
the other foot upon the Rhumb, which ſtretcheth towards the
ſaid ſecond Cape; and moving the two Compaſſes (ſo opened)
by theſe two Rhumbs off from the Land, the very ſame Point
where the two feet which came from the two Capes do meet,
you may affirm to be the very Point where your Ship is. And
then meaſuring by the degrees of the Equinoctial, you may ſee
what diſtance there is from the ſaid Point to either of the foreſaid Capes, or to any other place, which you think good, for it
is a very eaſie matter, if you know the point where your Ship is.
PROB.

PROB. XIX.

Of Tides, *and how by help of the* Globe *you may in general judge of them.*

Ivide the Equinoctial into 30 equal parts, as was directed in Prob. 54. of the laft Book. Thefe 30. equal parts reprefent the 30. daies of the Moons Age.

Then on the North and South point of the Compafs in the outmoft Verge of the Horizon, Write with red Ink 12. From the North Eaftward, *viz.* at the Point North and by Eaft, Write 11¼. At the next point to that the fame way, *viz.* North North Eaft, Write 10½. At the next, *viz.* North Eaft and by North, Write 9¾. And fo forward to every point of the Compafs; rebating of the laft hour ¾ till you come to 12. in the South; where you muft begin again to mark that Semi-Circle alfo in the fame order you did the laft. In this Circle is then reprefented the Points of the Compafs the Sun and Moon paffeth by every Day; and the Figures annexed reprefent the twice 12. hours of Day and Night.

Having thus prepared your Globe and Horizon, you may by having the Moons Age, and the point of the Compafs on which the Moon maketh full Sea at any Place given, find at what Hour of Day or Night it fhall be high Tide in the fame Place. Thus,

It is a known Rule that a North and South Moon makes high water at *Margarate.* Therefore Bring the firft point of ♈ to the North or South point in the Horizon, and Elevate the North Pole into the *Zenith* : Then count in the Equinoctial the Daies of the Moons Age numbred in red figures ; and the Hour and minutes (written in red figures annexed to the names of the Windes) that ftands againft the Moons Age fhall be the Hour of High Tide on that Day or Night at *Margarate.*

The End of the Third Book.

R

The Fourth BOOK,

Shewing the Practical Use of the

GLOBES:

Applying them to the Solution of

Aftrological Problemes.

P R Æ F A C E.

THe Practife of Aftrology *is grounded upon a two-fold Doctrine. The firft, for erecting a Figure of Heaven, placing the Planets in it, finding what Afpects they bear each other, and in what Places they are conftituted, &c. and this we call the Aftronomical part of A-ftrology.*

The fecond is, how to judge of the events of things by the Figure erected : and this is indeed the only Aftrological part.

The firft of thefe I fhall briefly handle; becaufe what therein is propofed may be performed by the Globe, both with fpeed, eafe, delight, and demonftration. The fecond I fhall not meddle with, but refer you to the whole Volumnes already written upon that Subject.

PROB.

PROB. I.

To Erect a Figure of the 12 Houfes *of* Heaven.

BEfore you erect a Figure of the 12 Houfes of Heaven it will be requifite you place the Planets ☊ and ☋, according to their Longitude and Latitude upon the Globe, as was directed in Prob. 55. of the fecond Book : for then, as you divide the Houfes of your Figure by the Circle of Pofition, you may by infpection behold in what Houfes the Planets are fcituated, and alfo fee what fixed Stars they are applying to, or feparating from.   But to the matter.

There is difagreement between the Ancient and Modern Aftrologers, about erecting a Figure of Heaven. M r *Palmer* in his Book of *Spherical Problemes* Chap. 48. mentions four feveral waies, and the Authors that ufed them; whereof one of them is called *the Rational way* ufed by *Regiomontanus*; and now generally practifed by all the Aftrologers of this Age. This way the face of Heaven is divided into twelve parts, which are called the twelve Houfes of Heaven numbred from the Afcendent or angle at Eaft downwards, with 1, 2, 3, &c.   As in the following Figure.

In a Direct Sphear, *viz.* under the Equator thefe twelve Houfes are twelve equal parts : but in an Oblique Sphear they are unequal parts, and that more or lefs according to the quantity of the Sphears obliquity.

Thefe twelve Houfes are divided by 12. Semi-Circles of Pofition; which are Semi-Circles paffing from the two interfections of the Horizon and Meridian through any Star, degree, or point in the Heavens.

Four of thefe Houfes are named Cardinals. The firft and moft eminent of thefe Cardinals is the firft Houfe, or the Angle of Eaft, called the Afcendent; where the Semi-Circle of Pofition is the fame with the Eaftern Semi-Circle of the Horizon. The fecond Cardinal is the tenth Houfe, or the Angle of South ; called *Medium Cæli,* or *Culmen Cæli* ; where the Semi-Circle of Pofition is the fame with the Semi-Circle of the Meridian above the Horizon. The third Cardinal is the feventh Houfe, or the Angle of Weft ; called the Defcendent; where the Semi-Circle of Pofition is the fame with the Weftern Semi-Circle of the Hori-

zon.

zon. The fourth Cardinal is the fourth Houſe, or Angle of North; called *Imum Cæli* ; where the Semi Circle of Poſition is the ſame with the Semi-Circle of the Meridian under the Horizon.

The degrees and minutes of the *Ecliptick* upon the Cuſps of theſe four Houſes (that is, upon the beginning of theſe Houſes) are found all at once only by bringing the Riſing degree of the *Ecliptick* to the Horizon: (for the Horizon repreſents the Cuſp of the Aſcendent:) and then ſhall the Meridian cut the degree of the *Ecliptick* on the Cuſp of the tenth Houſe. The Weſtern Semi-Circle of the Horizon ſhall cut the degree of the *Ecliptick* on the Cuſp of the Seventh Houſe : and the Semi-Circle of the Meridian under the Horizon ſhall cut the degree of the *Ecliptick* on the Cuſp of the fourth Houſe.

If you have the day of the Moneth, you may by Prob. 3. of
the

the fecond Book find the Suns Place; and if you have the Hour of the Day you may by firſt rectifying the Globe, as by Prob. 2. of the ſame Book, turn about the Globe till the Index of the Hour-Circle point to the ſame Hour in the Hour-Circle, and you will then at the Eaſtern Semi-Circle of the Horizon have the de-gree of the *Ecliptick* that is Riſing, and by Conſequence (as afore-ſaid) all the Cardinal points in their reſpective places.

Now to find what degree of the *Ecliptick* occupies the Cuſps of the other eight Houſes of Heaven ; Do thus, The Globe rectified, as aforeſaid, Move the Semi-Circle of Poſition up-wards till 30 degrees of the Equator ſhall be contained be-tween it and the Eaſtern Semi-Circle of the Horizon ; ſo ſhall the Semi-Circle of Poſition cut in the *Ecliptick* the degree and minute of the *Ecliptick* on the Cuſp of the twelfth Houſe; and its oppoſite degree and minute in the *Ecliptick* ſhall be the Cuſp of the ſixth Houſe, (for you muſt note that if you have but the de-gree and minute of the *Ecliptick* upon the Cuſps of ſix of the Houſes, the oppoſite degrees and minutes of the *Ecliptick* ſhall immediately poſſeſs the Cuſp of every oppoſite Houſe.)

Oppoſite de-grees and mi-nutes of the Ecliptik *poſ-ſeſs the* Cuſps *of oppoſite* Houſes.

Then move the Circle of Poſition over 30. degrees more of the Equinoctial, ſo ſhall the degree of the *Ecliptick* cut by the Circle of Poſition be the degree of the *Ecliptick* upon the Cuſp of the eleventh Houſe; and its oppoſite degree in the *Ecliptick* ſhall be upon the Cuſp of the fifth Houſe. The degree of the *Ecliptick* upon the Cuſp of the tenth and fourth Houſes was found as before. Then remove the Circle of Poſition to the Weſtern ſide of the Meridian, and let it fall towards the Horizon till 30. degrees of the Equator are contained between the Meri-dian and it, ſo ſhall the degree of the *Ecliptick* cut by the Semi-Circle of Poſition be the degree of the *Ecliptick* on the Cuſp of the Ninth Houſe; and the oppoſite degree of the *Ecliptick* ſhall be upon the Cuſp of the third Houſe. Let the Semi-Circle of Poſition fall yet lower, till it paſs over 30. degrees more of the Equator, ſo ſhall the degree of the *Ecliptick* cut by the Semi-Circle of Poſition be the degree of the *Ecliptick* on the Cuſp of the eighth Houſe ; and the oppoſite degree of the *Ecliptick* ſhall be upon the Cuſp of the ſecond Houſe. The degrees of the *Ecliptick* on the Cuſp of the ſeventh Houſe, and Aſcendent, were found as before.

Example.

Example.

I would erect a Figure of Heaven for *July* 27. 5. hours o. minutes Afternoon, 1658. in the Latitude of *London, viz.* 51 ½ degrees, North Latitude.

I firſt place the Planets, ♌, and ☋, on the Globe, as by Prob. 55. of the Second Book was directed: yet not exactly as I find them in the *Ephemeris*, for that ſhews only their place in the *Ecliptick* at Noon : Therefore I conſider how many degrees or minutes each Planets motion is in a whole Day or 24. Hours, by ſubſtracting the *Ecliptical* degrees and minutes of the Planets place that Day at Noon from the *Ecliptical* degrees and minutes of the Planets place the next Day at Noon : or contrarily if the Planet be Retrograde: for the remains of thoſe degrees and minutes is the motion of the Planet that Day ; Therefore proportionably to that motion I place the Planet forward in the *Ecliptick* (or backward if it be Retrograde:) As if the Sun ſhould move forward 1 degree, that is 60 minutes in a whole Day, or 24 Hours, then in 12 hours he ſhould move 30 minutes, in 6 hours 15 minutes, in 4 hours 10 minutes, in 1 hour 2 ½ minutes, and ſo proportionably for any other ſpace of Time: which I conſider before I place the Planets on the Globe.

Having thus placed the Planets on the Globe, I Elevate the North Pole 51 ½ degrees above the Horizon, and find the Suns place by Prob. 3. Book 2. to be in ♌ 14. degrees 9. minutes, Therefore I bring ♌ 14. degrees 9. minutes to the Meridian, and the Index of the Hour-Circle to 12. Then I turn the Globe Weſtwards, becauſe it is Afternoon, till the Index point to 5. Hours afternoon, and with a quill I faſten the Globe in this poſition : Then I examine what degree of the *Ecliptick* is at the Aſcendent or Horizon, and find ♐ 27. 47. to which Signe degree and minute Ⅱ 27. 47. is oppoſite, and therefore, as aforeſaid upon the Cuſp of the Seventh Houſe: Lifting up the Circle of Poſition till it paſs over 30 degrees of the Equator from the Horizon upwards I find ♐ 7. 5. cut by it in the *Ecliptick*, which is the Signe degree and minute upon the Cuſp of the twelfth Houſe, and its oppoſite Signe degree and minute is Ⅱ 7. 5. which is upon the Cuſp of the ſixth Houſe: Then lifting up the Circle of Poſition again till it paſs over 30 degrees more of the Equinoctial, I find

cut

cut by the Circle of Poſition ♏ 21. 18. which is the Signe degree and minute upon the Cuſp of the eleventh Houſe; and its oppoſite Signe degree and minute is ♉ 21. 18. which is upon the
Cuſp of the fifth Houſe: ♏ 3. 20. is at the Meridian, which is the
Cuſp of the tenth Houſe, and the Signe degree and minute oppoſite to it is ♉ 3. 20. which is on the Cuſp of the fourth Houſe.
Then taking the Semi-Circle of Poſition off its Poles, I place it
on the Weſt ſide the Meridian, and let it fall towards the Horizon
till it paſs over 30 degrees of the Equator from the Meridian, and
find the Circle of Poſition cut the *Ecliptick* in ♎ 1. 9. which is
the Signe degree and minute on the Cuſp of the ninth Houſe; oppoſite to ♎ 1. 9. is ♈ 1. 9. therefore ♈ 1. 9. is upon the Cuſp of
the third Houſe : Letting the Circle of Poſition fall yet lower till
it paſſes over 30 degrees more of the Equator, I find it cut the
Ecliptick in ♌ 6. 47. which is the Signe degree and minute up-

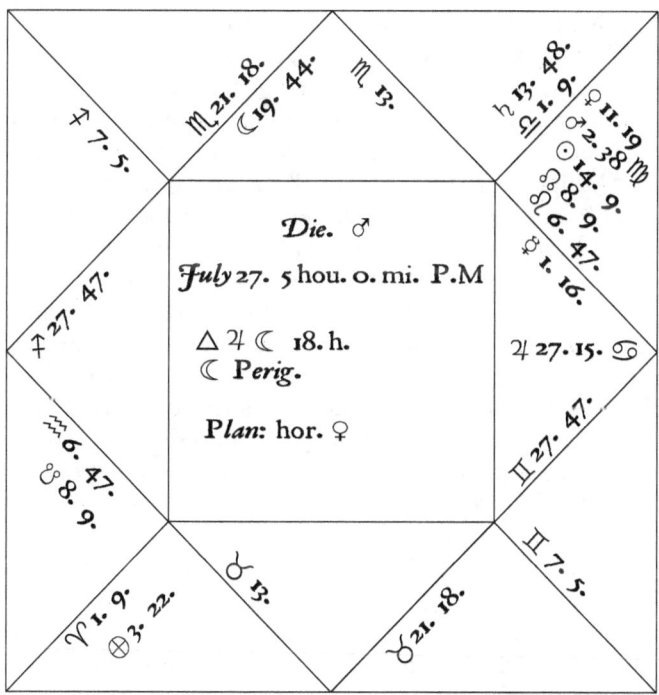

on the Cuſp of the eighth Houſe; and its oppoſite Signe degree and
minute is ♒ 6. 47. which is upon the Cuſp of the ſecond Houſe.
So have you a Figure of the Face of Heaven: which if you have
future uſe for, you may ſet down the ſeveral charracters in the
proper places of a Figure, as they are on the other ſide the leaf.

PROB. II.

To Erect a Figure of Heaven *according to* Campanus.

R *Egiomontanus* as aforeſaid makes the beginning of every
Houſe to be the Semi Circle drawn by the ſide of the Se-
mi Circle of Poſition according to the ſucceſſion of every 30ᵗʰ
degree of the Equator from the Horizon. But *Campanus* makes
it to be the Semi-Circle drawn by the ſide of the Semi-Circle of
Poſition according to the ſucceſſion of every 30ᵗʰ degree of the
Prime Verticle, or Eaſt *Azimuth;* which is repreſented by the
Quadrant of Altitude placed at the Eaſt point.

The four Cardinals are the ſame, both according to *Regiomontanus,* and *Campanus* : but the other eight Houſes differ : Therefore when you would find them according to *Campanus;* Rectifie
the Globe and Quadrant of Altitude, and bring the lower end of
the Quadrant of Altitude to the Eaſt point in the Horizon: Then
count from the Horizon upwards 30 degrees on the Quadrant of
Altitude, and bringing the Circle of Poſition to thoſe 30 degrees
examine where the Circle of Poſition cuts the *Ecliptick,* which at
the aforeſaid time is in ♏ 29. 40 for that degree and minute is
upon the Cuſp of the twelfth Houſe, and its oppoſite degree and
minute in the *Ecliptick viz.* ♉ 29. 40. is upon the Cuſp of the
ſixth Houſe : Lift up the Circle of Poſition 30 degrees higher
upon the Quadrant of Altitude (*viz.* to 60 degrees) and the
Circle of Poſition will cut the *Ecliptick* in ♏ 15. degrees for
the Cuſp of the eleventh Houſe, and its oppoſite degree and minute in the *Ecliptick viz.* ♉ 15. is upon the Cuſp of the fifth
Houſe. The degree and minute of the *Ecliptick* on the Cuſp of
the Tenth and Fourth Houſes is at the Meridian.

Then transfering the Circle of Poſition to the Weſt ſide of the
Meridian and the Quadrant of Altitude to the Weſt point in the
Horizon, Let the Semi-Circle of Poſition fall 30 degrees from the
Meridian on the Quadrant of Altitude, and it will cut in the *E-*
cliptick

cliptick ♎ 16 degrees, for the Cufp of the ninth Houfe, and its oppofite degree and minute in the *Ecliptick viz.* ♈ 16. is upon the
Cufp of the third Houfe : Let fall the Circle of Pofition 30 degrees lower on the Quadrant of Altitude, and it will cut the *E-
cliptick* in ♍ 2 degrees, for the Cufp of the eight Houfe, and its
oppofite degree *viz.* ♓ 2. degrees is on the Cufp of the fecond
Houfe : The Cufps of the Seventh and Afcendent is the fame
with *Regiomontanus viz.* ♊ 27. 47. and ♐ 27. 47. The Figure follows.

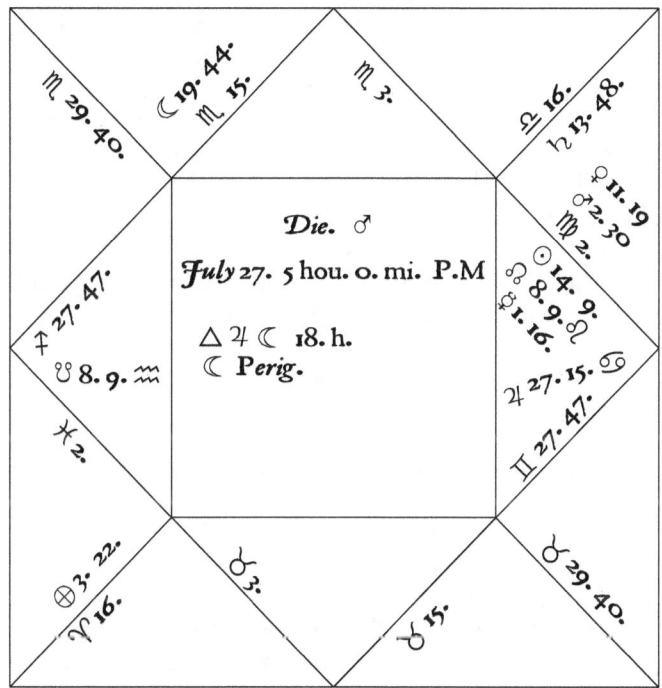

PROB. III.

To find the length of a Planetary Hour.

AStrologers divide the Artificial day (be it long or fhort)
into 12 equal parts, and the Night into 12 equal
parts : Thefe parts they call Planetary Hours. The

firſt of theſe Planetary Hours takes its denomination from the Planetary Day ; and the reſt are named orderly from that Planet according to the ſucceſſion of the Planetary Orbs : As if it be Munday that is, the Moons day, (as by Prob. 42. of the ſecond book) the Planet reigning the firſt Hour ſhall be ☾ , the Planet ruling the ſecond Hour ſhall be ♄, the third Planetary Hour ſhall be ♃, the fourth ♂, the fifth ☉, the ſixth ♀, the ſeventh ☿ : Then begin again with ☾ for the eight Planetary, ♄ for the ninth and ſo through the whole Day and Night, till the Sun Riſe again the next Day.

The length of this Planetary Hour is found by the Globe, thus : The Globe rectified ; Bring the Suns place to the Eaſt ſide the Horizon and make a prick at the degree of the Equator that comes to the Horizon with it. Then remove the Suns place to the Meridian, and count the number of degrees of the Equator comprehended between that prick and the degree now at the Horizon; and divide that number of degrees and minutes by 6. becauſe there is 6 Planetary Hours paſt ſince Noon; and the Quotient ſhall ſhew the number of degrees and minutes that paſs through the Meridian in one Planetary Hour.

Example.

July 27. 1658. I would know the length of the Planetary Hour here at *London* : I Rectifie the Globe, and bring the Suns place *viz.* ♌ 13. 50. to the Eaſtern ſide the Horizon and find 115 degrees of the Equator come to the Horizon with it; to this 115 degrees I make a prick : Then I turn the Suns place to the Meridian and find 226 degrees of the Equator at the Horizon. Therefore I either count the number of degrees between the pricks and the degree of the Equator at the horizon, or elſe ſubtract the leſſer from the greater but both waies I find 111 degrees of the Equator to paſs through the Meridian (or the Horizon) in ſix Planetary Hours. Therefore dividing 111. by 6. I find 18. degrees 30. minutes of the Equator to paſs through the Meridian in one Planetary Hour: which 18. degrees 30 minutes reduced into Time yeelds 72. minutes, by accounting for every 15. degrees one Hour for 1. degree 4. minutes, and for half a degree 2. minutes of Time and ſo proportionably ; ſo that the length of a Planetary Hour, *July* 27 is 1 common Hour and 14 minutes, here at *London*.   P R o B.

PROB. IV.

The length of a Planetary Hour *known, to find what* Planet *Reigneth any given* Hour *of the* Day, *or* Night.

THe Globe Rectified as in the laſt Probleme, Turn about the Globe till the Index of the Hour Circle points to the Hour of the Day in the Hour Circle. Then count the number of degrees comprehended between the degree of the Equator at the Horizon and the prick in the Equator, made as in the laſt Probleme, and reduce that number of degrees into minutes of Time, by rekoning 4. minutes of Time for every degree of the Equator. Reduce alſo the number of degrees and minutes that paſs through the Meridian in one Planetary Hour into minutes by allowing (as aforeſaid) 4. minutes for every degree, and then divide the firſt number by the ſecond and the Quotient ſhall be the number of Planetary Hours ſince Sun Riſing. Having the number of Planetary Hours ſince Sun Riſing, Reckon the firſt Planetary Hour by the name of that Planet that bears the denomination of the Day, the ſecond Planetary Hour by the Planet ſucceeding that in order, the third by the next in order and ſo for all the reſt till you come to the laſt Planet *viz.* ☽ ; and then begin again with ♄, and ſo to ♃ &c. till you have reckoned ſo many Planets as there are Planetary Hours ſince Morning : and that Planet the number ends on ſhall be the Planet Reigning that Planetary Hour.

Example.

July 27. 1658. aforeſaid, I would know what Planet Rules at 5 a clock paſt Noon : The length of the Planetary Hour this Day (found by the laſt Probleme) is 1. hour 14. minutes. Therefore the Globe Rectified, I bring the Index of the Hour Circle to the Hour of the Day *viz.* 5 a clock in the Hour-Circle, and then count the number of degrees between the Prick made, as by the laſt Probleme and the degree of the Equator at the Horizon; and find them 187. which I reduce into minutes, by allowing for every degree 4 minutes; and that gives 748 minutes. This 748 minutes I divide by the minutes contained in one Planeta-

S 2　　　　　　　　　　　　　　　　　　　ry

ry Hour this Day, *viz.* by 72. and find 10. hours 8. minutes; which ſhews there are 10. Planetary Hours and 8. minutes paſt and gon ſince Sun Riſing. Therefore ♂ being the Planet after whoſe name the Day is called *viz. Dia Martis,* ♂ is as aforeſaid, the Ruler of the firſt Planetary Hour : From him I count the Planet ſucceding, which is ☉ for the ſecond Hour; from ☉ I count the Planet ſucceding, which is ♀ for the third Hour, and ſo on to ☿, and ☾ : and then I begin the Round again with ♄, ♃, ♂, and ☉, till I come again to ♀, which is the tenth Planetary Hour ſince Sun Riſing : and the minutes remaining being 8. ſhews that there is 8. minutes paſt ſince ſhe began to Reign.

PROB. V.

To find Part *of* Fortune *by the* Globe.

Count the number of degrees and minutes contained between the Suns place and the Moons place, begining at the Suns place and counting according to the ſucceſſion of Signes till you come to the Moons place: and having found that number of degrees and minutes, add them to the number of degrees and minutes Aſcending, reckoned from the firſt point of ♈. If the ſum exceed 360. caſt away 360. and the remainder ſhall be the number of degrees and minutes from the firſt point in ♈, in which *Part of Fortune* falls. But if it do not exceed 360, you have already the number of degrees and minutes from the firſt point of ♈ in which you muſt place *Part of Fortune.*

Example.

I would find the place of *Part of Fortune* for the time of our Figure: I ſeek the two pricks repreſenting ☉ and ☾, and find ☉ in ♌ 14. 9. and ☾ in ♍ 19. 44. therefore counting from the Suns place to the Moons place according to the ſucceſſion of Signes, I find 95. degrees 35. minutes, contained between them: This 95. degrees 35. minutes I add to 267. degrees 47. minutes, the degree and minute contained between the firſt point of ♈ and the Aſcendent; and they make together 363. degrees 22. minutes. This exceeds 360. therefore I caſt away 360. and the remains are 3 degrees 22. minutes, for the place in the *E-cliptick* of *Part of Fortune,* reckoned from the firſt point of ♈.

Therefore

Therefore this character ⊕ which reprefents *Part of Fortune*, I fet in its proper place of the Figure, as I did the Planets.

PROB. VI.

To find in what Circle *of* Pofition *any* Star, *or any degree of the* Ecliptick *is*.

Circles of Pofition are numbred from the Horizon upwards, upon the Quadrant of Altitude placed at the Eaft or Weft point of the Horizon. Therefore when you would find what Circle of Pofition any Star or degree of the *Ecliptick* is in, Rectifie the Globe and Quadrant of Altitude, and bring the lower end of the Quadrant of Altitude to the Eaft or Weft point of the Horizon, and lift up the Circle of Pofition till it come to the Star or degree of the *Ecliptick* propofed : and the number of degrees the Circle of Pofition then cuts in the Quadrant of Altitude is the number of the Circle of Pofition that the Star or degree of the *Ecliptick* is in. If the Star or degree of the *Ecliptick* be under the Horizon, turn the Globe about till 180. degrees of the Equator pafs through the Meridian, then will the Star or degree of the *Ecliptick* be above the Horizon : Lift up then the Circle of Pofition (as before) to the Star or degree of the *Ecliptick* and the number of degrees of the Quadrant of Altitude the Circle of Pofition cuts on the Eaft fide, is the number of Circles of Pofition the Star was under the Horizon on the Weft fide : Or fo many degrees as the Circle of Pofition cuts on the Quadrant of Altitude in the Weft fide the Horizon is the number of the Circles of Pofition the Star or degree of the *Ecliptick* was under the Horizon on the Eaft fide.

PROB. VII.

To find the Right Afcenfions, *the* Oblique Afcenfions, *and the* Declinations *of the* Planets.

EXamine the Right Afcenfions and Declinations of thofe pricks made to reprefent each Planet, in Prob. 1. of this Book; and work by them as you were directed to work by the Sun, in Prob. 26, 27, 28. of the fecond Book.

S3   PROB.

PROB. VIII.

How to Direct *a* Figure, *by the* Globe.

TO Direct a Figure is to examine how many degrees of the Equinoctial are moved Eaſtwards or Weſtwards, while any Planet or Star in one Houſe comes to the Cuſp or any other point of any other Houſe.

When you would Direct any *Promittor* to any *Hylegiacal* point examine the degree of the Equator at the Meridian; then turn about the Globe till the *Promittor* come to the *Hylegiacal* point, and examine again the degree of the Equator at the Meridian : and by ſubſtracting the leſſer from the greater you will have the number of Degrees that paſſed through the Meridian whiles the place of the *Promittor* was brought to the *Hylegiacal* point: and that number of degrees ſhall be the Arch of Direction.

Example.

I would Direct the Body of the Moon in our Figure aforeſaid to *Medium Coeli,* or the tenth Houſe : I find by the Globe 203. degres 30. minutes of the Equator at the Meridian with the tenth Houſe and turning the Globe till the prick made to repreſent the Moon come to the Meridian. I find 227. degrees 20. minutes of the Equator come to the Meridian with it. Therefore I ſubſtract the leſſer from the greater *viz.* 203. degrees 30. minutes from 227. degrees 20. minutes, and have remaining 23 degrees 50 minutes.

This 23. degrees 50. minutes ſhews that 23. Years 10. Moneths muſt expire ere the Effects promiſed by the Moons preſent poſition ſhall opperate upon the ſignification of the tenth Houſe.

If the Body of the Moon had been Directed to any other point than the Meridian or Horizon; you muſt have Elevated the Circle of Poſition to the point propoſed; and have under-propped it to that Elevation, and then have turned about the Globe till the prick repreſenting the Moon, had come to the Circle of Poſition; and then the degrees of the Equator that ſhould have paſſed

through

through the Meridian whiles this motion was making, ſhould be the number of degrees of Direction; and ſignifie in Time as fore-ſaid.

PROB. IX.

Of Revolutions : *and how they are found by the* Globe.

BY Revolution is meant the Annual Converſion of the Sun to the ſame place he was in at the Radix of any Buſineſs.

When you would find a Revolution by the Globe, firſt find the Right Aſcenſion of Mid Heaven at the Radix of the Buſineſs, as by Prob. 26. of the ſecond Book you were directed to find the Right Aſcenſion of the Sun; and to it add 87 degrees for eve-ry Year ſince the Radix : Then ſubſtract 360 ſo oft as you can from the whole and the Remains ſhall be the Right Aſcenſion of Mid Heaven for the Annual Revolution.

If you count the number of degrees of the Equator contained between the Right Acenſion of the Mid Heaven and the Right Aſcenſion of the Sun, and convert that number of degrees into Time by allowing for every 15. degrees 1 Hour of Time it will ſhew, if the Suns place be on the Weſtern ſide of the Meridian the number of Hours and minutes Afternoon the Revolution ſhall happen on, but if on the Eaſt ſide the Meridian, the number of Hours and minutes Before noon the Revolution ſhall happen on.

PROB. X.

How a Figure of Heaven *may be erected by the* Revoluti-on *thus found.*

SEek the degree of Right Aſcenſion of Mid Heaven, and bring it to the Meridian, ſo ſhall the four Cardinal points of the Globe be the ſame with the four Cardinal points in Heaven at the time of the Revolution. The other Houſes are found by the Circle of Poſition : as in the firſt Pro-bleme of this Book.

The End of the Fourth Book.

The Fifth BOOK,

Shewing the Practical Use of the

GLOBES:

Applying them to the Solution of

Gnomonical Problemes.

PRÆFACE.

Dyals are of two sorts, Pendent, and Fixed. Pendent are such as are hung by the hand, and turned towards the Sun; that by its Beams darting through smal Pin-holes made for that purpose, the hour of the Day may be found. These are of two sorts, Vniversal, and Particular.

Vniversal Dyals *are those commonly called Equinoctial or Ring-Dyals : They are used by Sea-men and Travellers, that often shift Latitudes.*

Particular *are such as are made and only serve for Particular Latitudes. Of these sorts are the several Dyals discribed on Quadrants, Cilinders, &c.*

Fixed Dyals *shall be the matter of this discourse; and they are such as are made upon fixed Planes, and shew the Hour of the Day by a* Stile *or* Gnomon *made Parallel to the Axis of the World.*

Of

Of the ſeveral Kinds of Dyal Plains : and how you may know them.

A Plain in Dyalling is that flat whereon a Dyal is diſcribed.

There is ſome diſagreement among Older and Later Authors in the naming of Plains: for ſome name them according to the Great Circle in Heaven they ly in : and others according to the ſcituation of the Poles of the Plains. Thus they which name them according to the Great Circle in Heaven their Plains ly in, call that an Horizontal Plain, which others call a Vertical Plain; thoſe Vertical, which others will call Horizontal; and thoſe Polar, which others call Equinoctial.

However they be called it matters not, ſo you can but diſtinguiſh their kinds, which with a little conſideration you may eaſily learn to do : For remembring but upon what grounds either the Older or Later Authors gave the Plains their Names, upon the ſame grounds you may alſo learn to know them. I confeſs both waies admit of ſome juſt exception againſt for in the Older Rule a Plain about the Pole, is called an Equinoctial Plain; when as to a ſudden apprehenſion it would ſound more ſignificant to call it a Polar Plain, as Later Authors do : Again, Later Authors call an Horizontall Plain a Vertical Plain; when as it ſounds more ſignificant to call it an Horizontal Plain, as Older Authors do becauſe it lie flat upon the Horizon : But I ſhall give you the names according to both Rules, and leave you to your liberty to accept of which you pleaſe.

Firſt therefore, you have an Equinoctial Plain otherwiſe called a Polar Plain. This Plain hath two Faces, upper, and under : Theſe two Faces ly in the Plain of the Equinoctial : the upper Face beholding the Elevated Pole, the under Face the depreſſed Pole.

2. An Horizontal Plain, otherwiſe called a Vertical Plain : it lies in the Plain of the Horizon, directly beholding the *Zenith*.

Erect Plains, otherwiſe called Horizontal Plains are the ſides of Walls, and theſe are of ſeven ſorts, *viz* 1. Erect Direct Vertical, North or South, 2. Erect Direct, Eaſt or Weſt. 3. E r e ct Vertical Declining. 4. Erect Inclining Direct. 5. Erect Inclining Declining. 6. Erect Reclining Direct. 7. Erect Reclining Declining.

3. Erect Vertical, North or South Direct, otherwiſe called

T   Direct

Direct North or South Horizontals, behold the North or South
Directly, and ly in the Eaſt or Weſt *Azimuth*.

4. Erect Direct Eaſt or Weſt, otherwiſe called Direct Eaſt or
Weſt Equinoctials, behold the Eaſt or Weſt Directly, and lies in
the Plain of the Meridian, having its Poles in the Equinoctial.

5. Erect Vertical Declining Plains, otherwiſe called Declining Horizontals, do not behold the North or South Directly, but
ſwerves from them ſo much as the *Azimuth* Parallel to their
Plains ſwerves or Declines from them.

6. Erect Inclining Direct Plains, have the upper ſide of their
Plains Inclining or coming towards you, and their Plains do exactly behold either the Eaſt, Weſt, North, or South.

7. Erect Reclining Direct Plains, have the upper ſide of their
Plains Reclining or falling from you, and their Plains exactly beholding either the Eaſt, Weſt, North, or South.

8. Erect Reclining Declining, or Erect Inclining Declining
Plains are thoſe Plains which are either Inclining or Reclining,
but do not behold the Eaſt, Weſt, North or South, Directly but
ſwerve or Decline more or leſs from them.

9. Polar Plains are Parallel to the Axis of the World, and to
the Meridians that cuts the Eaſt and Weſt, or North and South,
points of the Horizon.

All theſe kinds of Plains have two Faces ; the one beholding
the North Pole with the ſame reſpect that the other beholds the
South Pole; except the Equinoctial Plain, which, becauſe neither
Pole is Elevated, hath but one Face : yet that one contains as
many Hour lines as two other Faces.

Theſe two Faces or Plains will receive juſt 24. hour lines, for
the 24 Hour-lines of Day and Night: for ſo much as the one ſide
or Face wanteth or exceedeth 12. the other ſide ſhall either exceed or want of 12.

Every Dyal Plain is Parallel to the Horizon of ſome Country
or other in the World: therefore a Dyal made for any Horizon
in the World may be ſet to ſuch a Poſition that it will ſhew
you the Hour of the Day in your own Habitation : At leaſt for
ſo long as the Sun continues upon that Plane.

All Plains may be aptly demonſtrated by the Globe, by ſetting it correſpondent to all the Circles in Heaven, as by Prob. 2.
of the ſecond Book : for if you imagine the Globe in that Poſition were preſt flat into the Plain of any Circle, that Flat ſhall
                                                      repreſent

reprefent a Dyal plain, which fhall be called after the name of that Circle it is preft into.

Thus if the Quadrant of Altitude be applyed to any degree of *Azimuth,* and you imagine the Globe were preft flat to the edge of the Quadrant of Altitude, fo much as that *Azimuth* Declines from the Eaft, Weft, North, or South, in the Horizon, fo much fhall that flat on the Globe be faid to Decline either from the Eaft, Weft, North, or South. Or if you imagine the Globe were preft flat down even with the Plain of the Horizon, that flat fhall reprefent an Horizontal Plain; becaufe as was faid before, the Plain lies in that Circle cal'd the Horizon.

The *Style* or *Gnomon* is that ftraight wyre that cafts the fhadow upon the Hour of the Day: it is alwaies placed Parallel to the Axis of the World.

There are feveral waies to find the fcituation of all Plains; but the readieft and fpeedieft is by a Clinatory. The Clinatory is made of a fquare board, as A B C D, of a good thicknefs, and the larger the better; between two of the fides is difcribed on the Center A a Quadrant as E F divided into 90 equal parts or degrees, which are figured with 10, 20, 30, to 90 ; and then back again with the Complements of the fame numbers to 90 : between the *Limb* and the two Semidiameters is made a Round Box, into which a Magnetical Needle is fitted ; and a Card of the Sea Compafs, divided into 4 Nineties, beginning their numbers at the Eaft, Weft, North, and South points of the Compafs, from which points the oppofite fides of the Clinatory receives their Names of Eaft, Weft, North, or South. Upon the Center A whereon the Quadrant was difcribed is faftned a Plumb-line, having a Plumbet of Lead or Brafs faftned to the end of it, which Plumb-line is of fuch length that the Plumbet may fall juft into the Grove G H below the Quadrant, which is for that purpofe made of fuch a depth that the Plumbet may ride freely within it, without ftopping at the fides of it. See the Figure annexed.

With this Clinatory you may examine the fcituation of Plains. As if your Plain be Horizontal ; it is Direct : and then for the true fcituating your Dyal you have only the true North and South line to find: which is done only with fetting the Clinatory flat down upon the Plain, and turning it towards the right or left hand, till you can bring the North point of the Needle to hang

T 2   juft

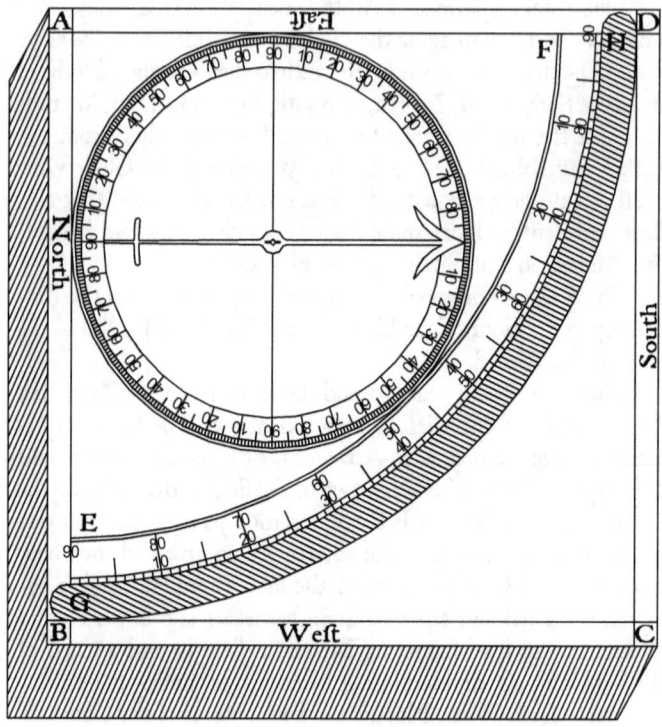

juſt over the Flower-de-luce: for then if you draw a line by either
of the ſides Parallel to the Needle, that line ſhall be a North and
South line. But herein reſpect muſt be had to the Variation of
the Compaſs in the Place you make your Dyal: for if the North
point of the Needle ſwerves from the North point of the World,
then have you not a true North and South line. But if in your
Place there be no Variation of the North point of the Needle
from the North point of the World (as now it happens here at
London) then the line drawn by the ſide of the Clinatory (as a-
foreſaid) ſhall be a true North and South line.

But admit there be Variation, Having by Prob. 19. of the
third Book found the number of degrees of this Variation towards the Eaſt or Weſt, count the ſame number of degrees from
the

the North point in the Card either to the Eaſtwards or Weſtwards, and note the degree in the Card terminating at that number, for that degree ſhall be the North point ; and its oppoſite
degree the South point: 90. degrees from it either way ſhall be
the Eaſt and Weſt points.

Therefore, whereas before you were directed to turn the
Clinatory till the North point of the Needle point to the Flower-
de-luce on the Card you muſt now turn (or move) the Clinatory till the North point of the Needle hang juſt over the degree
of Variation thus found; and then a line drawn as aforeſaid, by the
ſide of the Clinatory Parallel to the Needle ſhall be a North and
South line or (to ſpeak more properly) a Meridional line.

You may find a Meridian line ſeveral other waies ; as firſt;
If the Sun ſhine juſt at Noon, hold up a Plumb-line ſo as the
ſhadow of it may fall upon your Plain; and that ſhadow ſhall be
a Meridian line.

Secondly, on the backſide the Clinatory diſcribe a Circle, and
draw a line through the Center to both ſides the Circumference;
croſs this line with an other line at Rght Angles in the Center,
ſo ſhall the Circle be divided into four equal parts. Theſe four
parts you muſt mark with Eaſt, Weſt, North, South, and divide each
of them into 90. degrees. In the Center of this Plain erect a
ſtraight wyer prependicularly: when you would find a Meridian
line examine by the tenth Prob. of the ſecond Book the Amplitude of the Suns Riſing or Setting from the Eaſt or Weſt points,
and waiting the juſt Riſing or Setting that Day, turn the Inſtrument about till the ſhadow of the wyer falls upon the ſame degree from the Eaſt or Weſt the Amplitude is of, for then the
North and South line in the Inſtrument will be the ſame with the
North and South line in Heaven.

Thirdly by the Suns *Azimuth* : Find the *Azimuth* of the
Sun by Prob. 22. of the ſecond Book : and at the ſame inſtant
turn the Inſtrument till the ſhadow of the wyer fall upon the degree on the Inſtrument oppoſite to the degree of the Suns *Azimuth,* ſo ſhall the Meridional line of the Inſtrument agree with
the Meridional line in Heaven.

You may the ſame way work by the *Azimuth* of any Star.
Only, whereas the ſhadow of the wyer ſhould fall upon the oppoſite degree aforeſaid : Now you muſt place a Sight or Perpendicular upon that oppoſite degree, and turn the Inſtrument a-

bout till the wyer at the Center, the Sight in the oppofite degree of the Stars *Azimuth,* and the Star in Heaven, come into one ftraight line, fo fhall the Meridian line of the Inftrument agree with the Meridional line in Heaven.

Fourthly It may be found by any Star obferved in the Meridian, if two Perpendiculars be erected in the Meridian line of your Inftrument; for then by turning the Inftrument till the two Perpendiculars and the Star come into a ftraight line, the Meridian line of your Inftrument will be the fame with the Meridian line in Heaven. See more waies in Mr. *Palmer* on the *Planifphear* Book 4. Chap. 9.

If your Plain either Recline or Incline, apply one of the fides of your Clinatory Parallel to one of the Semi-diameters of the Quadrant to the Plain, in fuch fort that the Plumb-line hanging at liberty may fall upon the Circumference of the Quadrant, for then the number of degrees of the Quadrant comprehended between the fide of the Quadrant Parallel to the Plain, and the Plumb-line fhall be the number of degrees of Reclination, if the Center of the Quadrant points upwards, or Inclination if the Center points downwards.

If your Reclining or Inclining Plain Decline, draw upon it a line Parallel to the Horizon, which you may do by applying the back-fide of the Clinatory, and raifing or depreffing the Center of the Quadrant till the Plumb-line hang juft upon one of the Semi-diameters, for then you may by the upper fide of the Clinatory draw an Horizontal line if the Plain Incline, or by the under fide if it Recline. If it neither Incline or Recline, you may draw an Horizontal line both by the upper and under fides of the Clinatory. Having drawn the Horizontal line, apply the North fide of the Clinatory to it, and if the North end of the Needle points directly towards the Plain, it is then a South Plain. If the North point of the Needle points directly from the Plain, it is a North plain: but if it points towards the Eaft, it is an Eaft Plain: if towards the Weft a Weft Plain. If it do not point directly either Eaft, Weft, North, or South, then fo many degrees as the Needle declines from any of thefe four points to any of the other of thefe four points, fo many degrees is the Declination of the Plain, with refpect (as aforefaid) had to the Variation of the Compafs.

Or if you find the *Azimuth* of the Sun by its Altitude obferved juft when its beams are coming on or going off your

Plain

Plain, that *Azimuth* ſhall be the *Azimuth* of your Plain.

Or you may erect a wyer Perpendicularly on your Plain, and wait till the ſhadow of that wyer comes to be Perpendicular with the Horizon, which you may examine by applying a Plumb-line to it, for then the ſhadow of the Plumb-line and the ſhadow of the Perpendicular will be in one: then taking the Altitude of the Sun you may by Prob. 22. of the ſecond Book find its *Azimuth*, and thereby know in what *Azimuth* the Plain of your Dyal lies: for the *Azimuth* your Plain lies in is diſtant from the *Azimuth* of the Sun juſt 90. degrees.

PROB. I.

How by one poſition of the Globe *to find the diſtances of the* Hour-lines *on all manner of* Plains.

YOu may have Meridian lines drawn from Pole to Pole through every 15. degrees of the Equinoctial, to repreſent the Horary motion of the Sun both Day and Night; and when the Pole of the Globe is Elevated to the height of the Pole in any Place and one of theſe Meridian lines be brought to the Brazen Meridian, all the reſt of the Meridian lines ſhall cut any Circle which you intend ſhall repreſent the Plain of a Dyal in the number of degrees on the ſame Circle that each reſpective Hour-line is diſtant from the Noon-line point in the ſame Circle.

Thus if you ſhould enquire the diſtance of the Hour-lines upon an Horizontal Plain in *Londons* Latitude; The Pole of the Globe (as aforeſaid) muſt be Elevated 51 ½ degrees, and one of the Meridian lines (you may chuſe the Vernal Colure) be brought to the Brazen Meridian, which being done, you are only to examine in the Horizon (Becauſe it is an Horizontal Plain) at what diſtance from the Meridian (which in Horizontals is the Noonline) the ſeveral Meridians drawn on the Globe interſect the Horizon, for that diſtance in degrees ſhall be the diſtance on a Circle divided into 360. degrees that each reſpective Hour-line muſt have from the Meridian or a Noon line choſen in the ſame Circle; and lines drawn from the Center of that Circle through thoſe degrees ſhall be the Hour lines of an Horizontal Plain.

If it be an Erect Direct South Dyal you enquire after; Keeping

ing the Globe in its former poſition, apply the Quadrant of Altitude to the *Zenith,* and its lower end to the Eaſt point of the
Horizon, for then (as was ſhewed in the *Preface*) by imagining the
Globe to be preſt flat to the graduated edge of the Quadrant of
Altitude, that flat ſhall be a South Plain and the number of degrees the Meridians cuts in the Quadrant of Altitude numbred
from the *Zenith* downwards ſhall be the number of degrees
that each Hour line ſhall be diſtant from the Meridian or Noon-
line in a Circle of 360. degrees, and lines drawn from the Center of that Circle through thoſe degrees ſhall be the Hour lines
of half the Day: the Hour lines for the other half of the Day are
of the ſame diſtance from the Noon-line, with theſe; only they
muſt be placed on the other ſide the Noon line.

    If your Plain be not Direct but declines Eaſt or Weſt you
muſt number the Declination Eaſtwards or Weſtwards reſ-
pectively in the degrees of the Horizon and (the Quadrant of
Altitude

Altitude fcrewed to the *Zenith,* (as aforefaid) bring the lower
end of the Quadrant of Altitude to the faid degrees of Declination, and the number of degrees cut by the Meridians in the
Quadrant of Altitude numbred downwards, is the number of
degrees that the Hour-lines are diftant from the Noon line in a
Circle of 360. degrees : And lines drawn from the Center of
that Circle through thofe degrees be the Hour lines of half the
Day. And if you turn about the Quadrant of Altitude upon
the *Zenith* point till the lower end of it come to the degree of
the Horizon oppofite to the degree of Declination found before,
the Meridian lines on the Globe (as before) fhall cut the Quadrant of Altitude in the number of degrees (counted downward)
that each Hour-line is diftant from the other fide the Noon-line :
And lines drawn from the Center of that Circle through thofe
degrees fhall be the Hour-lines of the other half of the Day,

   If your Plane Decline, and alfo Recline or Incline, you muft ufe
the *Gnomonical* Semi-Circle, difcribed in Prob. 12. which muft be
Elevated on the Quadrant of Altitude when it is fet to the Declination (as by the former Rule) according to the complement of
Reclination, or Inclination : But if your Plane be Direct, and Recline, or Incline, it muft be fet to the Meridian, and the Meridians
on the Globe fhall cut that Semi Circle in the number of degrees
counted from the Quadrant of Altitude if the Plane Declines, or
from the Brafen Meridian, if it be Direct, that the feveral Hour
lines are diftant from a line Perpendicular to an Horizontal line, in
a Circle divided into 360 degrees ; And lines drawn from the
Center through thofe degrees fhall be the Hour-lines of fuch
Reclining or Inclining Planes.

   If your Plane be an Eaft or Weft, either Direct or Declining; or
an Equinoctial Plane (for they are upon the matter all one) you
may better conceive how they are to be made, then make them
by the Globe. And for the help of your fancy herein, take M^r
Blagraves conceit, who in his Book 6. Chap. 8. very properly
demonftrates the Rules for projecting the Hour-lines on thefe
Planes. He propofes to take 12. wyers bowed into exact Circles, all of equal Diameter, and fet together at equal diftance
one from another in two oppofite points, as in two Poles, and to
have a ftraight line to pafs from one Pole to another, as an Axis.
Thefe 12 Wyers fhall reprefent 24 Meridional Semi-Circles,
Or indeed they may reprefent the Globe it felf, containing 24 Me-
    ridional

ridional Semi-Circles to be
diſcribed on the Globe, as
aforeſaid; And if you place
the Horizon of the Globe
Horizontal, and the North
and South points of the
Globe towards the North
and South points in Hea-
ven, and bring one of theſe
Wyer Meridians directly
under the Braſen Meridian,
and the Axis of this Wyer-
Globe in the Plain of the
Horizon, and faſten a Thred
in the middle of the Axis,
that thred drawn from the
middle of the Axis by
every one of theſe wyers
ſhall, if prolonged till it
touch an Eaſt and Weſt
line drawn directly under
or over the points *Zenith*
or *Nadir,* point out on that
Eaſt and Weſt line the diſtances of each Hour-line

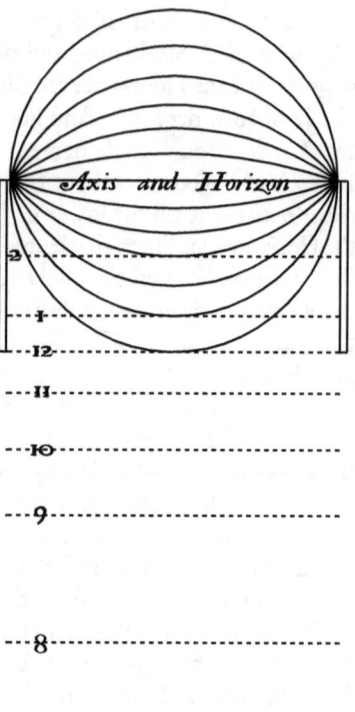

from the 12 a clock line ; And lines drawn at Right Angles
through that Eaſt and Weſt line, ſhall be the Hour lines of an
Eaſt or Weſt Plane, or of an Equinoctial Plane.

    The moving this thred from wyer to wyer repreſents the mo-
tion of the Sun, which as it paſſes over all the Meridians cauſes
the ſhadow of that Meridional Semi Circle which it is directly
over, and the Axis, and the Meridional Semi-Circle directly oppoſite to the upper Meridional Semi-Circle to fall all into one
ſtraight line : And upon what point in the Eaſt and Weſt line
(mentioned before) that ſhadow-line ſhall fall is marked out by
the application of the thred as aforeſaid : and is an Hour-line on
any of the foreſaid Planes.

    If you underſtand this Probleme rightly, you do already know
how to draw the Hour lines upon all manner of Planes, and need
no further Inſtructions; yet partly fearing a raw Student ſhould

not

not clearly underſtand theſe Rules, and partly doubting (becauſe other Authors have been more Copious upon this Subject) that I ſhould be cenſured to be too ſparing of my pains, if I ſhould lightly touch ſo eminent a Doctrine as Dyalling is : Therefore I ſhall more diſtinctly handle Dyalling by the Globe, according to the way or Method that other Authors have uſed, and that after ſo plain a manner as poſſibly my Genius can deviſe.

PROB. II.

To make an Equinoctial Dyal.

Deſcribe a Circle, on a ſquare board or Plane as B C E D, and through A the Center thereof draw a ſtraight line Parallel to one of the ſides, as B E ; Croſs that ſtraight line with another ſtraight line as C D at Right Angles, ſo ſhall the Circle be divided into 4 equal parts : Divide each of theſe four equal parts into 90. degrees ; as in the Figure. This Circle ſhall repreſent the Horizon.

Erect a wyer exactly perpendicular to the Center of the Plane; and that wyer ſhall be the *Gnomon* or *Style* of the Dyal.

Then Elevate one of the Poles of your Globe into the *Zenith,* and bring the Equinoctial Colure to the Meridian. And becauſe in every hours Time 15 degrees of the Equator paſſes through the Meridian in Heaven, therefore turn the Globe till 15 degrees of the Equator paſs through the Meridian of your Globe; ſo ſhall the Colure paſs by 15 degrees of the Horizon alſo. Therefore from the Center of your Plane draw ſtraight lines through 15. degrees from one of the Semidiameters both waies : and thoſe ſtraight lines ſhall be two Hour-lines : Then turn the Globe till 15 degrees more of the Equator paſs through the Meridian, and you will find as before, the Colure paſs by 15 degrees more of the Horizon: therefore on your Plane number 15. degrees further beyond both the former lines, and from the Center draw ſtraight lines through both thoſe 15. degrees, and they ſhall be two Hour lines more. For all the other Hour lines turn the Globe till 15. degrees of the Equator at a time paſs through the Meridian, as before, and you will find that for every 15. degrees of the Equator that paſſes through the Meridian, the Colure will paſs through 15. degrees of the Horizon : therefore thoſe

Hour

Hour lines muſt be drawn from the Center according to the ſucceſſion of every 15 degrees on your Plane. Having drawn the Hour lines, you may ſet figures to them, beginning to number your Hour lines from one of the Diameters, marking it with XII, and the next Hour line to the left hand with I, and the next II, the next III, &c. to XII. and begin again with I, II, III, &c. till you come to the other XII, where you began : and then your Dyal is finiſhed.   See the Figure.

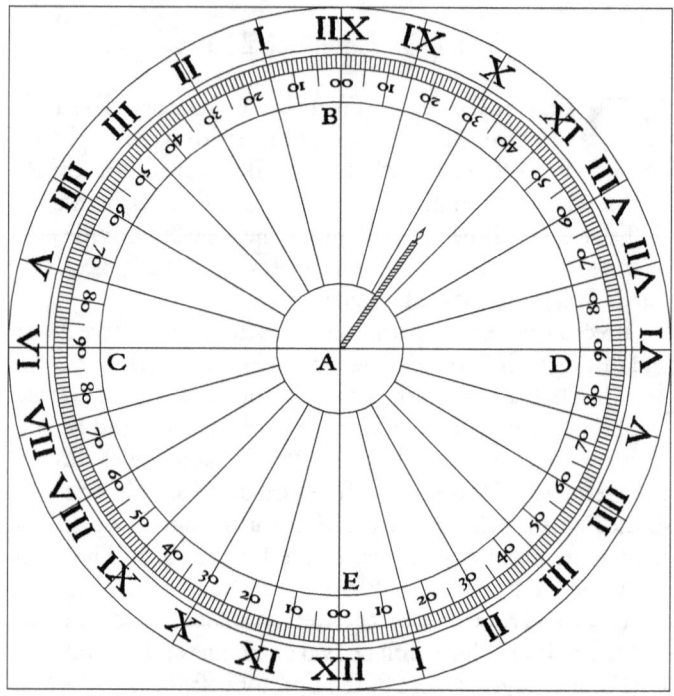

This is an Univerſal Dyal, and ſerves in all Latitudes: therefore when you place it you muſt ſet one of the XIIˢ downwards, and the Axis Parallel to the Axis of the World.

But note, Both faces of this Dyal ought to be divided, and the *Gnomon* muſt appear on both ſides like the ſtick in a Whirligig, which childeren uſe; or elſe you muſt turn it upſide down, ſo oft as the Sun paſſes the Equinoctial.

PROB.

PROB. III.

To make an Horizontal Dyal.

Dﬁcribe a Circle on your Plane, as C B D E, and through
the Center A of that Circle draw a Meridian line, as
B E; croſs that line at Right angles with another line,
as C D; ſo ſhall your Circle be divided into four e-
qual parts : Divide each of theſe four parts into 90. degrees ;
ſo ſhall the whole be divided into 360. Theſe 360 degrees repreſent the 360 degrees of the Horizon, which a Meridian line
drawn through the place of the Sun runs through in every 24.
Hours : The motion of which Meridian line through the degrees of the Horizon is Regular in a Parallel Sphear ; for in e-
qual Time it moves an equal Space throughout the whole Circle, *viz.* it will paſs through 15. degrees of the Horizon in one
Hours Time, (or which is all one) whiles 15. degrees of the E-
quator paſſes through the Meridian ; as was ſhewed in the laſt
Probleme : But in an Oblique Sphear its motion through the
Horizon is Irregular, and that more or leſs according to the
more or leſs Obliquity of the Sphear : For far Northwards or
Southwards you may ſee this Meridian line paſs through 40, 50,
yea 60. degrees of the Horizon in one Hours time, *viz.* whiles
15. degrees of the Equator paſſes through the Meridian : but
in an other Hours time you will ſcarce have 4 or 5 degrees paſs
through the Horizon whiles 15 degrees of the Equator paſſes
through the Meridian.

But that you may know the motion of the Sun (repreſented
by this Meridian line) through the Horizon in all Latitudes ; E-
levate the Pole to the Elevation of your Place, and chuſe inſtead
of a Meridian line drawn through the Place of the Sun the Vernal Colure to be your Meridian line ; both becauſe it is moſt
viſible ; and becauſe from thence the degrees of the Equator are
begun to be numbred, ſo that what ſo ever decimal degree of
the Equator you light on at the Meridian, or elſe where, you
will find its number from that Colure already ſet down to your
hand, without either adding to, or ſubſtracting from it. Bring
this Colure therefore to the Meridian, and the Index of the
Hour Circle to 12. in the Hour Circle. Then turn the Globe

Weſtwards, and ſo oft as 15 degrees of the Equator paſſes through the Meridian, ſo oft you muſt examine what degrees of the Horizon the Vernal Colure cuts; and thoſe degrees and minutes ſo cut by the Vernal Colure muſt be found in the Circle C B D E, beginning your account or reckoning at B towards D, and markt with Pricks : through which Pricks you muſt draw lines from the Center A, and thoſe lines ſhall be the Hour lines after noon. Then bring the Colure to the Meridian again to find the Fore-noon Hour-lines, and turn the Globe Eaſtwards, and ſo oft as 15 degrees of the Equator paſſes through the Meridian, ſo oft you muſt examine what degrees of the Horizon the Vernal Colure cuts ; and thoſe degrees and minutes ſo cut by the Vernal Colure muſt be found in the Circle C B D E, begining your reckoning from B towards C, and markt with Pricks : through which Pricks you muſt draw lines from the Center A, and thoſe lines ſhall be the Fore-noon Hour-lines.

The ſe Hour-lines muſt be markt from the Meridian line, *viz.* the line A B, which is the 12 a clock line towards D, with I, II, III &c. till you have numbred to the Hour of Sun ſet (found by Prob. 7. of the ſecond Book) the longeſt Day, and from the Meridian line towards C with XI, X. IX, &c. till you have numbred to Sun Riſing the longeſt Day.

The *Stile* muſt be placed in the Center and Elevated ſo many degrees above the Plane, as the Pole is elevated above the Horizon of the Place.

Example of the whole.

I would make an Horizontal Dyal for *Londons* Latitude: Therefore I Elevate the North Pole 51 ½ degrees above the Horizon, and bring the Vernal Colure to the Meridian, and the Index of the Hour Circle to 12 on the Hour Circle;

And turning { 1 } a clock or till 15 { 11. 4 }
the   Globe { 2 } deg of the Equa- { 24. 15 }
Weſtwards, { 3 } tor paſs through { 38. 4 } from the Meridian.
till the In- { 4 } the Meridian ; I { 53. 36 }
dex points { 5 } find the Colure { 71. 6 }
to   { 6 } cut the Horiz. in { 90. }

The ſe are the diſtances of the Hour lines from Noon till *6* at Night : and to theſe diſtances on the Plane (counting from B towards

wards D,) I make pricks ; and from the Center I draw lines
through thefe Pricks; and thefe lines are the Hour lines from 12
to 6 Afternoon. But the Sun in the longeft Day fhines till paft 8
at Night, as you may find by Prob. 48. of the fecond Book,
therefore here wants the two Evening Hour lines; which though
they may be found after the fame way I found the former, (*viz.*
by continuing the turning of the Globe Weftwards) yet that I
may the fooner reduce my work to the Plane I Count the number of degrees between the 6 a clock line and the 5 a clock line
in the Circle on the Plane; for the fame number of degrees counted
from D towards E is the diftance of the 7 a clock Hour line from
the 6 a clock Hour line; and the number of degrees contained between the 6 a clock Hour line and the 4 a clock Hour line is the
diftance of the 8 a clock Hour line from the 6 a clock Hour-line.

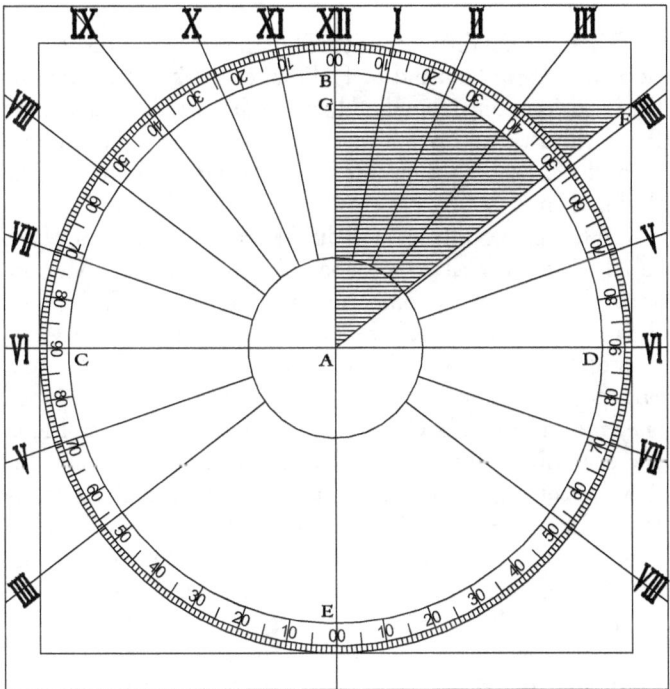

Or I need not draw the 7 and 8 a clock Hour lines, till I have
drawn the forenoon Hour lines: for then by laying the edge of a
                                                            Ruler

Ruler (that will reach through the oppofite fide of the Plane) to the Morning 7 and 8 a clock Hour lines, I may by the fide of that Ruler draw lines from the Center through the oppofite fide of the Plane, and thofe lines fhall be the 7 and 8 a clock Hour lines Afternoon.

Having thus all the Afternoon Hour lines, I bring the Vernal Colure to the Meridian again ; fo fhall the Index again point to 12. Therefore, as before I turned the Globe Weftwards, fo now

turning it Eaftwards, till the Index points to

11	a clock, or till 15	11.40
10	deg. of the Equa-	24.15
9	tor pafs through	38. 4
8	the Meridian, I	53.36
7	find the Colure	71. 6
6	cut the Hori. in	90.

from the Meridian.

Thefe are the diftances of the Hour lines from Noon to 6. a clock in the Morning: and thefe diftances I feek in the Circle of the Plain (counting from the Noon line B towards C) and mark them with Pricks; through which pricks (as before) I draw lines from the Center to the outfide the Plane: and thofe lines fhall be the Hour lines.

Or having the diftance of all the Afternoon Hour-lines, I have alfo the diftance of all the forenoon Hour lines from the Meridian; as you may fee by comparing the two former Tables. For the 1 a clock Hour line Afternoon is equidiftant from the Meridian or Noon line with the 11 a clock Hour line before Noon, *viz.* they are both 11 degrees 40 minutes diftant from the Noon line, and the 2 a clock Hour line Afternoon is from the Noon line equidiftant with the 10 a clock Hour line Beforenoon; for they are both 24. degrees 15. minutes diftant from the Meridian or Noon line: and fo all the other Morning Hour lines are diftant from the Noon line by the fame fpace that the fame number of Afternoon Hour lines (told from the Meridian on the contrary fide the Noon line) are diftant from the Meridian.

Whence it follows, that fince (as aforefaid) the fame number of Hour lines after 6 at Night, and before 6 in the Morning have the fame diftance from the 6 a clock line that the fame number of Hour lines before 6 at Night and after 6 in the Morning have from the 6 a clock line ; and fince the fame number of Hour lines

lines before Noon are equidiftant from the Meridian or Noon line by the fame fpace of degrees that the fame number of Hour lines Afternoon are; It follows (I fay) that having found the diftance of the fix Hour lines either before or after Noon, you have alfo given the diftance of all the other Hour lines.

If you will have the half Hour lines placed on your Dyal you muft turn the Globe till the Index points to every half Hour in the Hour Circle, as well as to the whole, and examine the degrees of the Horizon cut by the Vernal Colure, as you did for the whole Hours; and in like manner transfer them to your Plane.

Having thus drawn all the Hour lines I count from the Noon line 51½ degrees, the Elevation of the Pole here at *London*; and from the Center A I draw a ftraight line, as A F through thefe 51½ degrees, for the *Gnomon* or *Style,* and prolong it to the fartheft extent of the Plane : From this *Gnomon* or *Style* I let fall a Perpendicular upon the Noon line, as F G: (this Perpendicular is called the Subftile, and this Perpendicular and its Bafe (which is the Noon line) *and Hypothenufa* (which is the *Gnomon*) fhall make a Triangle, which being erected upon the Bafe, fo as the Subftile may ftand Perpendicular to the Plane, the *Hypothenufa* A F fhall be the *Gnomon,* and be Parallel to the Axis of the World; and caft a fhadow upon the Hour of the Day.

PROB. IIII.

To make an Erect Direct South Dyal.

DRaw on your Plane an Horizontal line as C A D, as was fhewed in the Preface : in the middle of this line (as at A) difcribe as on a Center the Semi-Circle C B D : from the Center A let fall a Perpendicular, which fhall divide the Semi-Circle into two Quadrants each of which Quadrants you muft divide into 90 degrees. Then Rectifie the Globe, Quadrant of Altitude, Colure and Hour Index : thus, Elevate the Pole of the Globe to the Latitude of your Place, and fcrew the Quadrant of Altitude to the *Zenith,* Then bring the Vernal Colure to the Meridian, and the Index of the Hour Circle to the Hour of 12. in the Hour Circle; fo fhall your Globe, Quadrant of Altitude, Colure and Hour Index be Rectified. And thus you muft alwaies Rectifie them for the making of moft forts of Dyals by the

X   Globe.

Globe. Then to make an Erect Direct South Dyal, Bring the
lower end of the Quadrant of Altitude to the Weft point of the
Horizon; And turn the Globe Weftwards till the Index points
to all the Hours Afternoon ; and examine in what numbers of
degrees from the *Zenith* the Colare cuts the Quadrant of Altitude when the Index points to each Hour: for a line drawn from
the Center A through the fame number of degrees reckoned
from the Perpendicular A B (which is the 12 a clock line) towards D on the Plane, fhall be the fame Hour lines the Index
points at.

Thus in our Latitude, *viz.* 51½ degrees, the Vernal Coloure being brought to the Meridian and the Index to 12 ;

If you turn the ⎧ 1 ⎫ a clock, or till 15 deg. ⎧ 9. 18 ⎫
Globe Weft- ⎪ 2 ⎪ of the Equator pafs ⎪ 19. 15 ⎪
wards, till the ⎨ 3 ⎬ through the Meridi- ⎨ 32. 5 ⎬ counted from
Index points ⎪ 4 ⎪ an, the Colure will ⎪ 48. 0 ⎪ the *Zenith.*
to ⎪ 5 ⎪ cut the Quadrant of ⎪ 67. 4 ⎪
⎩ 6 ⎭ Altitude in ⎩ 90. ⎭

And thefe are the diftances of the Afternoon Hour lines; which
you muft transfer to the Eaft fide of your Plane, *viz,* from B
towards D; and draw lines from the Center A through thefe di
-ftances; and thefe lines fhall be your Afternoon Hour lines.

Note (once for all) when the Colure goes off that Circle you
examine the Hour diftances in, the Sun will fhine no longer upon
that Plane; As in this example the Colure goes off the Quadrant of Altitude at 6 a clock, therefore the Sun will not fhine
longer then till 6 a clock upon this Plane.

The Hour lines before Noon have the fame diftance from the
Meridian that the Afternoon Hour lines have, as was fhewed in
the laft Probleme: Only they muft be drawn on the Weft fide
the Noon line, and counted from B towards C.

Otherwife.

You may reduce all Verticals into Horizontals; if you Elevate
the Pole of the Globe to the Complement of the Latitude of
your Place, and bring the Vernal Colure to the Meridian under
the Horizon, and the Index of the Hour Circle to 12; and turn
the

the Globe Weftwards; for as the Index paffes through every
Hour on the Hour Circle, the Colure fhews in the Horizon the
diftance of the feveral Afternoon Hour lines from the Meridian,
or 12 a clock line, in the Circle on your Plane, numbred from B
to D: and lines drawn from the Center through thefe diftances on
your Plane fhall be the Afternoon Hour lines of your Dyal.

Example.

Londons Latitude is 51 ½ degrees, Its Complement to 90. is
38 ½ . Therefore I Elevate the Pole 38 ½ degrees above the Horizon, and bring the Vernal Colure to the Meridian under the
Horizon, and the Index of the Hour Circle to 12 on the Hour
Circle. Then

Turning the ⎧ 1 ⎫ a clock, or till 15 ⎧ 9 18 ⎫ from the Interfecti-
Globe Weft- ⎪ 2 ⎪ deg. of the Equa- ⎪ 19 15 ⎪ on of the Meridian
wards, till the ⎨ 3 ⎬ tor pafs through ⎨ 32 5 ⎬ and the Horizon: as
Index of the ⎪ 4 ⎪ the Meridian, I ⎪ 48 0 ⎪ in the former Ta-
Hour Circle ⎪ 5 ⎪ find the Colure ⎪ 67 0 ⎪ ble.
points to ⎩ 6 ⎭ cut the Horizon in ⎩ 90 ⎭

    And thefe are the diftances of the 6 Hour lines from the Merid.

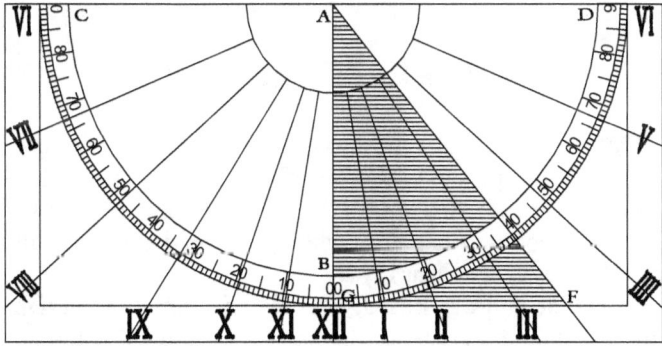

By this Example you may fee that it is eafie to reduce Verticals
into Horizonals : and Horizontals into Verticals : for this Erect
Direct South Dyal is an Horizontal Dyal to thofe People that
Inhabite 90 degrees from us, *viz.* in the South Latitude of 38 ½
degrees.
    Then make a Triangle, whereof the Noon line fhall be
                            X 2   Bafe :

Bafe: from it count the Complement of the Poles Elevation, *viz.* 38½ degrees, and through them draw the line A F, from the Center A which fhall be *Hypotenufa;* Then let fall a Perpendicular upon the Noon line A B, fo is your Triangle made. If this Triangle be erected Perpendicularly upon the Bafe or Noon line, The *Hypotenufa* A F fhall ftand Parallel to the Axis of the World, and caft a fhadow upon the Hour of the Day.

PROB. V.

To make an Erect Direct North Dyal.

IF the Erect Direct South Dyal were turned towards the North; and the line C A D were turned downwards, and the line marked with 7 be now marked with 5, and the line 8 with 4, the line 5 with 7, and the line 4 with 8, then have you of it a North Erect Direct Dyal.

All the other Hour lines in this Dyal are ufelefs, becaufe the Sun in our Latitude fhines on a North Face the longeft Day only before 6 in the Morning, and after 6 at Night.

PROB. VI.

To make an Erect Direct Eaft Dyal.

THefe forts of Dyals may better be demonftrated then made by the Globe; unlefs the Axis of your Globe were acceffible, as in the Wyer-Globe, fpecified in Prob. 1. Therefore when you would make an Eaft, or Weft Dyal, or a Polar Dyal.

Provide a fquare Board, as A B C D, draw the ftraight line *e f* upon it Parallel to the fides A C, and B D. and juft in the middle between them : Crofs this ftraight line at Right Angles with another ftraight line, as *g h*, quite through the Board.

Upon this Board with a little Pitch or Wax faften the Semi-Circle of Pofition, fo as both the Poles thereof may ly in the line *g h*, and the middle of the Semi-Circle marked *co* may ly upon the line *e f*, fo fhall *i* be the Center of the Semi-circle of Pofition:

In

In this Center make a fmal hole through the Board fit to receive
a Wyer or a Nail. So may you with this Circle of Pofition
thus fitted, and the fide C D applyed to a line of Contingence elevated to the Height of the Equinoctial, draw lines from the Center through every 15 degrees of the Circle of Pofition, and by
continuing them interfect the line of Contigence in the points
from whence the Hour lines of an Eaft or Weft Dyal is to be
drawn.

Example.

I would make an Erect Direct Eaft Dyal for *Londons* Latitude. Therefore I faften a Plumb-line a little above the place

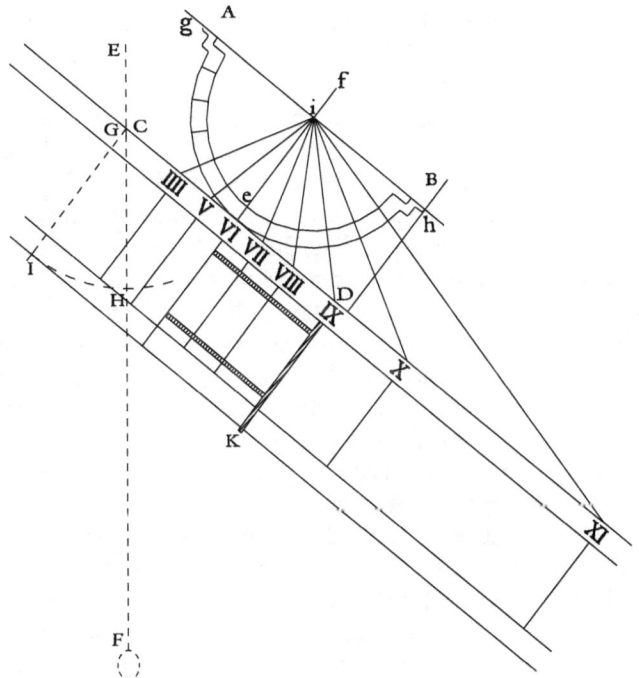

on the Wall where I intend to make my Dyal, and wait till it
hangs quietly before the Wall : then if the line be rub'd with
Chalk (like a Carpenters line) I may by holding the Plumbet
end clofe to the Wall, and ftraining it pretty ftif, ftrike with it a

      ftraight

ſtraight line as *Carpenters* do. This line ſhall be a Perpendicular as E F : I chuſe a convenient point in this Perpendicular, as
at G, for a Center ; whereon I diſcribe an occult Arch as H I :
this Arch muſt contain the number of degrees of the Elevation
of the Equinoctial counted between H and I, (which in our Latitude is 38½,) Therefore in a Quadrant of the ſame Radius with
the occult Arch I meaſure 38½ degrees, and ſet them off in the
Plane from H in the Perpendicular to I : Then from I to the
Center G in the Perpendicular, I draw the prickt line I G, and
this line ſhall repreſent the Axis of the World : I croſs this
Axis at Right Angles with the line G K and draw it from G to
K, ſo long as I poſſibly can: this line ſhall be the Contingent-line.
I find a convenient place in this Contingent line as, at VI, to
which I apply the ſide of the Board CeD, ſo as that the point *e*
may ly juſt upon VI in the Contingent line ; And having a thred
faſtned in the Center of the Semi-Circle of Poſition, I draw that
thred ſtraight over the firſt 15. degrees of the Circle of Poſition,
numbred from o towards *h*, and where the thred cuts the Contingent line I make a mark, for that mark ſhall be the mark for the
7 a clock line. From thence I remove the thred to 30. degrees
of the Semi-Circle, and draw it through the Contingent line, and
where it cuts the Contingent line, there ſhall be the mark for the
8 a clock line. From thence I remove the thred to 45. degrees
of the Semi-Circle and draw it through the Contingent line, and
where it cuts the Contingent line there ſhall be the mark for the
9 a clock line. From thence in like manner I remove the thred to
60. and 75. and where the thred cuts the Contingent line ſhall
be the mark for 10 and 11 a clock lines : The 12 a clock line
cannot be drawn on this Plane, as you may ſee, if you apply the
thred to 90 degrees, for though you ſhould draw it out never ſo
far yet would it never touch the Contingent line: becauſe it is
Parallel to the line *g h*, and lines Parallel never meet.

    But becauſe in our Latitude the Sun Riſes before 4. in the
Morning, therefore two Hour-lines are yet wanting, *viz.* 5. and 4.
which I may find either by applying the thred firſt to 15. and
next to 30 degrees from o towards *g* in the Semi-Circle, and ſo
marking where it cuts the Contingent line, as before : Or elſe by
transfering the diſtance of the ſame number of Hour lines from
the 6 a clock line already drawn on the ſide *e h* to the ſide *e g*,
as in Prob. 2. of this Book is more fully ſhewed.

Having

Having thus marked out on the Contingent line the diſtances of each Hour; I draw a line Parallel to the Contingent line, and draw lines from every Hour markt on the Contingent to croſs the Contingent line at Right Angles and continue each line to the line Parallel to the Contingent; and theſe lines ſhall be the Hour lines of an Eaſt Plane. To theſe Hour-lines I ſet Figures as in the Scheam may be ſeen.

The *Style* D K of this Dyal (as well as of others) muſt ſtand Parallel to the Axis of the World: it muſt be alſo Parallel to all the Hour lines, and ſtand directly over the 6 a clock line, and that ſo high as is the diſtance between the Center of the Semi-Circle of Poſition and the point where the 6 a clock line cuts the Contingent line: Or (which is all one) at ſuch a height as when it is laid flat down upon the Plane it may juſt reach the 3 a clock line.

PROB. VII.

To make an Erect Direct Weſt Dyal.

AN Erect Direct Weſt Dyal is the ſame in all reſpects with an Erect Direct Eaſt Dyal ; Only as the Eaſt ſhews the Fore-noon Hours, the Weſt ſhews the After noon Hours.

Thus if you ſhould draw the Eaſt Dyal on any tranſparent Plane, as on Glaſs, Horn, or an Oyled Paper, on the one ſide will appear an Eaſt Dyal, and on the other a Weſt. Only the Figures as was ſaid before (muſt be changed); for that which in the Eaſt Dyal is 11, in the Weſt muſt be 1 : that which in the Eaſt Dyal is 10, in the Weſt muſt be 2: that which in the Eaſt Dyal is 9 in the Weſt muſt be 3. &c.

PROB. VIII.

To make a Polar Dyal.

POlar Dyals are Horizontal Dyals under the Equinoctial : They are of the ſame kind with Eaſt and Weſt Dyals; Only whereas Eaſt and Weſt Dyals have but the Hour lines of half the longeſt Day diſcribed on them, theſe have all the Hour lines

of

of the whole Day ; and are marked on both ſides the Noon line :
as in the following Figure.

The *Style* of this Dyal muſt ſtand over the Noon line,
Parallel to the Plane; for then it will alſo be Parallel to the Axis

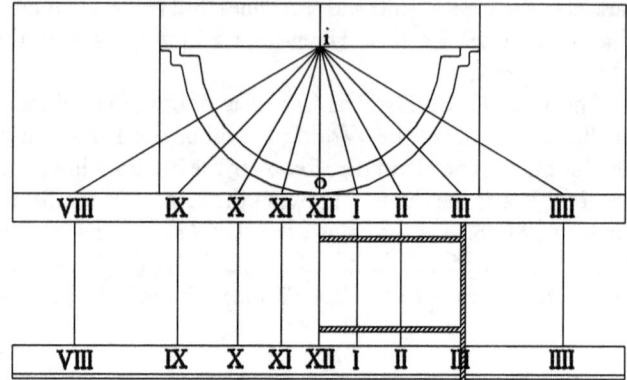

of the World : and its height above the Plane muſt be the diſtance between the Center *i* of the Semi-Circle and the point in
the Contingent line cut by the Noon-line. But I have inſerted the
Figure, which alone is ſufficient Inſtructions.

PROB. IX.

To make Erect South Dyals, Declining *Eaſtwards, or*
Weſtwards.

D Raw on your Plane an Horizontal line, and on it diſcribe
a Semi-Circle, as you were taught in Prob 4.

Then Rectifie the Globe, Quadrant of Altitude, Colure and
Hour Index, as by the ſame Probleme : and bring the lower
end of the Quadrant of Altitude to the degree of Declination
from the Eaſt or Weſt point, according as your Declination is
Eaſtwards or Weſtwards; for then the Quadrant of Altitude
ſhall repreſent a Plane declining from the South Eaſtwards, or
Weſtwards accordingly. Then turn the Globe Eaſtwards, till
the Index of the Hour-Circle points to all the Hours before
Noon, and examine in what number of degrees from the *Zenith*
the Colure cuts the Quadrant of Altitude, when the Index points
to each Hour, For a line drawn from the Center A through the
                                                         ſame

fame number of degrees reckoned from the Perpendicular A B, which is the 12 a clock line towards C on the Plane, fhall be the fame Hour-lines the Index points at.

Example.

I would make an Erect Dyal declining from the South towards the Eaft 27. degrees: The Globe, Quadrant of Altitude, Vernal Colure, and Hour Index Rectified, as before, I bring the lower end of the Quadrant of Altitude to 27. degrees counted from the Eaft point of the Horizon towards the North : Then

I turn the Globe Eaftwards till the Index points to				
	11	a clock, or till 15.	9. 43	
	10	deg. of the Equa-	19. 0	
	9	tor pafs through	25.57	counted from
	8	the Meridian, and	35. 10	the *Zenith*.
	7	find the Colure	45. 56	
	6	cut the Quadrant	60. 15	
	5	of Altitude in	79. 45	

And thefe are the diftances of the Fore-noon Hour-lines, which I feek in the Weft fide of the Plane, *viz.* from B towards C; and through thefe diftances I draw lines from the Center, and thefe lines fhall be the Fore-noon Hour-lines.

Now herein is a difference between Declining Dyals, and Direct Dyals: For having found the diftances of the Hour lines for one half of the Day, be it either for Before Noon or After Noon in a Direct Dyal, you have alfo found the diftances for the other half Day; becaufe, as was faid Prob. 3. Equal number of Hours have equal diftance from the Noon line : But in Declining Dyals it is not fo : Becaufe the Sun remaining longer upon that fide of the Plane which it declines to, then it doth upon the contrary fide, there will be a greater number of Hour lines upon it, and by confequence the diftance of the Hour lines lefs then on the contrary fide of the Plane.

Therefore for finding the After Noon Hour lines, I turn about the Quadrant of Altitude upon the *Zenith* point till the lower end of it come to the degree of the Horizon oppofite to that degree of Declination that the Quadrant of Altitude was placed at when I fought the Fore Noon Hour lines, *viz.* to 27. degrees

Y counted

counted rom the Weſt towards the South, and bring the Vernal Colure again to the Meridian, and the Index (as before) to 12. Then,

turning the Globe		a clock, or till 15 degr. of the		
turning the Globe	1	a clock, or till 15 degr. of the	11. 20	
Weſtwards till	2	Equator paſs through the Me-	26. 47	counted
the Index points	3	ridian, I find the Colure cut the	49. 20	from the
to	4	Quadrant of Altitude in	75. 52	*Zenith.*

And theſe are the diſtances of the After Noon Hour lines; which diſtances I ſeek in the Eaſt ſide of the Plane, *viz.* from B towards D (as before) and ſo drawing lines from the Center A through theſe diſtances, I have all the Afternoon Hour lines alſo drawn on my Plane.

You may note, that this Plane is capable to receive no more Hour lines After Noon then 4. for when the Colure goes off the Quadrant of Altitude, the Sun goes off theſe kind of Planes.

To theſe Hour lines I ſet their numbers, as you may ſee in the Figure.

Then to find both the diſtances of the *Subſtilar line* from the 12 a clock line, and the Elevation of the *Style* above the Plane, Bring the Colure to the number of degrees of the Planes Declination, counted in the Horizon from the South point towards the Eaſt point, and the Quadrant of Altitude to the degrees of the Planes Declination, counted in the Horizon from the Eaſt point towards the North, ſo ſhall the Quadrant of Altitude and the Colure cut each other at Right Angles ; and the number of degrees comprehended between the Colure and the *Zenith* in the Quadrant of Altitude, ſhall be the number of degrees between the *Subſtyler line* and the 12. a clock line, which in this Example is 19. degrees 45. minutes : And the number of degrees comprehended between the Quadrant of Altitude and the Pole, counted in the Colure, ſhall be the number of degrees that the *Style* is to be Elevated above the Plane; which in this Example is 33. degrees 40. minutes. Wherefore for the diſtance of the *Subſtyler line* from the 12 a clock line, I count in the Circle from the 12 a clock line in the contrary ſide of the Plane, *viz.* in the Weſt ſide, becauſe the Plane declines towards the Eaſt 19. degrees 45. minutes, as at D, and through that number of degrees and minutes from the Center A, I draw the line A G, which ſhall be the *Subſtyler line* : And from the *Subſtylar line* (either way) I number 33. degrees 40. minutes, the Elevation of the *Style* above the Plane, and
through

through thofe degrees and minutes I draw from the Center A, the line A F, for the *Style* or *Gnomon* ; Then I let fall the Per-pendicular F G upon the *Subftyle* A G : So is there a Triangle

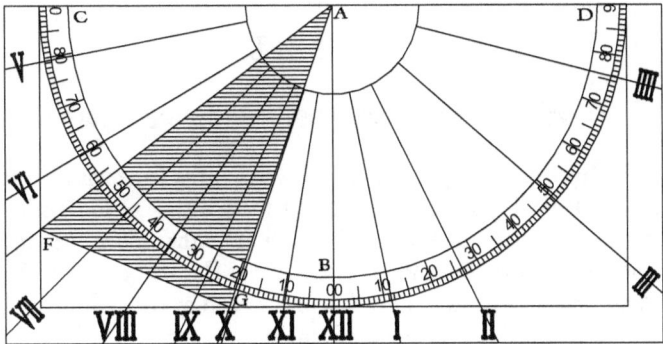

made, which if it be erected Perpendicularly upon the *Subftyle* A G, the *Style* A F fhall be Parallel to the Axis of the World, and caft a fhadow upon the Hour of the Day.

Here you may fee that in Declining Dyals the *Style* doth not ftand at the fame Elevation above the Plane, that it doth in Erect Direct Dyals; neither doth it ftand over the 12 a clock line; but fwerves from it towards the Quarter of Declination.

PROB. X.

To make a North Erect Dyal *declining* Eaftwards, *or* Weftwards.

AS in Prob. 5. an Erect Direct North Dyal hath the fame Delineation that an Erect Direct South Dyal hath, and differs only in the placing the Figures of the Hour lines : So a North Erect Dyal that declines Eaftwards, or Weftwards, differs from a South Erect Dyal that Declines Eaftwards, or Weftwards, the fame number of degrees, only in placing the Hour lines at the fame diftance on the contrary fide of the Plane, and by tranfpofing the Figures of 11 for 1 : 10 for 2 : 9 for 3. &c.

Thus, if you draw upon Glafs, Horn, or an Oyled Paper, the South Dyal Declining Eaftwards, as in the foregoing Probleme,

Y 2   and

and place it to its due ſcituation, the back ſide of it ſhall be a
North Dyal declining towards the Weſt ſo many degrees as the
foreſide Declines towards the Eaſt ; and the only difference in it
will be the Figures of the Hour lines ; as was ſaid before.

PROB. XI.

To make Direct Reclining, *or* Inclining Dyals.

Direct Reclining or Inclining Dyals are the ſame with Erect
Direct Dyals that are made for the Latitude of ſome other
Places : The Latitude of which Places are either more then the
Latitude of your own Place, if the Plane Recline, or leſs if the
Plane Incline; and that in ſuch a proportion as the arch of Re-
clination or Inclination of your Plane is.

Thus a Direct South Dyal Reclining 10. degrees in *Londons*
Latitude, *viz.* 51$\frac{1}{2}$ degrees, is an Erect Direct Dyal made for
the Latitude of 61$\frac{1}{2}$ degrees: And a Direct South Dyal Inclining 10. degrees in the Latitude 51$\frac{1}{2}$ degrees is an Erect Direct
Dyal in the Latitude of 41$\frac{1}{2}$ degrees : and is to be made according to the Directions in Prob. 4.

PROB. XII

To make Declining Reclining, *or* Declining Inclining Dyals.

The diſtances of the Hour lines either for a Declining
Reclining Plane, or a Declining Inclining Plane may
moſt eaſily be found upon the Plane of the Horizon.
That is (as ſome Authors call it) by the Horizontal Dyal,
by changing the Circles of the Globe one into another : So as
the Plane of the Horizon may ſerve to repreſent the Dyal Plane;
Yet this way not being natural, becauſe you muſt admit one Circle to be another, and that in Young Learners might ſometimes
breed a little difficulty, *Gemma Friſius*, *Metius*, and *Blaew* hath
preſcribed a thin Braſs plate to be made equal to a Semi-Circle of
the Equinoctial, and divided from the middle point of it either
way into 90 degrees, which may not unproperly be called a *Gnomonical Semi-Circle.* This Semi-Circle muſt be bowed cloſe

to

to the Body of the Globe into a Semi-Circular form, and so set
to any Reclination, or Inclination, and then it will represent a Reclining or Inclining Plane : And by the motion of the Colure
through the several degrees of this Semi-Circle the distances of
the Hour lines may be found : Thus,

The Globe, Quadrant of Altitude, Colure, and Hour Index,
Rectified; as by Prob. 4. Bring the lower end of the Quadrant
of Altitude to the degree in the Horizon of the Planes Declination, if your Plane be a South Declining Recliner, and count on the
Quadrant of Altitude from the *Zenith* downwards the number
of degrees of Reclination, or Inclination, and to that number of
degrees bring the middle of the *Gnomonical Semi-Circle,* and let
the ends of it cut the Horizon on either side in the degrees of the
Planes *Azimuth,* so shall the *Gnomonical Semi-Circle* represent
a Reclining Plane. And so oft as 15. degrees of the Equator
passes through the Meridian, so oft shall you enquire what degrees of the *Gnomonical Semi-Circle* the Colure cuts; for so many degrees asunder must the several respective Hour lines of a
Reclining Declining Plane be in a Semi-Circle divided into 180.
degrees.

But if your Plane be a South Declining Recliner, or a North
Declining Incliner; Bring the Quadrant of Altitude to the degree
of the Horizon opposite to the degree of the Planes Declination,
(because the upper side of the Plane lies beyond the *Zenith*)
counted from the South point in the South Recliners, and from
the North point in North Incliners.

Then find the heigth of the *Style,* and place of the *Substyle* :
thus, Keep your *Gnomonical Semi-Circle* in its position : But
turn the Quadrant of Altitude about on the *Zenith* point till the
lower end of it comes to the degree of the Horizon opposite to
the degree it was placed at before, and turn about the Globe till
the Colure cut the Quadrant of Altitude above the Horizon in
the number of degrees the Plane Reclines from the *Zenith*; so
shall the Colure cut the *Gnomonical Semi-Circle* at Right Angles; Then count the degrees contained between the middle of the
Gnomonical Semi-Circle and the Colure, for that number of degrees is the distance of the *Substyle* from a Perpendicular line in
the middle of your Plane, and must be placed Westwards of the
said Perpendicular, if your Plane decline from the South Eastwards; or Eastwards, if your Plane decline from the South West-

wards. Then obſerve how many degrees are contained between the Semi-Circle and the Pole; for that number of degrees is the number of degrees that the *Style* is to be Elevated above the *Subſtyle*.

Example.

Here at *London* I would make a Dyal upon a Plane Declining from the South Eaſtwards 30. degrees, and Reclining from the *Zenith* 20. degrees; *Londons* Latitude is 51 ½ degrees : Therefore, Having on the Plane diſcribed a Semi Circle, &c. as was directed Prob. 4. I Rectifie the Globe, Quadrant of Altitude, Colure, and Hour Index, as by the ſame Probleme and bring the lower end of the Quadrant of Altitude to 30. degrees from the North point of the Horizon towards the Weſt, becauſe that is the degree oppoſite to the degree of the Planes Declination, *viz.* to 30 degrees from the South Eaſtwards, And I bring the middle of the *Gnomonical Semi Circle* to 20. degrees of the Quadrant of Altitude counted from the *Zenith* downwards towards the Horizon, and the ends of the *Gnomonical Semi Circle* to the degrees of *Azimuth* the Plane lies in in the Horizon, *viz.* to 30. degrees from the Eaſt point Northwards, and to 30. degrees from the Weſt point Southwards, ſo ſhall 11. degrees 10. minutes of the *Gnomonical Semi Circle* be comprehended between the Quadrant of Altitude and the Braſen Meridian : Theſe 11. degrees 10. minutes ſhews that the 12 a clock line is diſtant from the Perpendicular A B 11. degrees 10. minutes : and becauſe the Plane Declines to the Eaſtwards, therefore the 12 a clock line muſt ſtand on the Weſt ſide the Plane 11. degrees 10. minutes. Then to find all the Fore Noon Hour lines,

I turn the Globe Eaſtwards till the Index points to				
	11	a clock, or till 15.	15. 8	
	10	degr. of the Equa-	18. 56	counted from the middle of the *Gnomonical Semi Circle.*
	9	tor paſs through	22. 37	
	8	the Meridian, and	26. 52	
	7	find the Colure	32. 37	
	6	cut the *Gnomonical*	42. 5	
	5	*Semi-Circle* in	62. 43	

And theſe are the diſtances of the Fore Noon Hour lines ; to which

which diftances you may fet Pricks on the Weft fide the Semi Circle of the Plane, *viz.* from B to C.

   The After Noon Hour lines are found by bringing the Colure again to the Meridian, and the Index of the Hour Circle to 12. for then

turning the	⌈ 1 ⌉	a clock, or till 15 degr. of the	⌈ 5. 45 ⌉	counted from
Globe Weft-	⎨ 2 ⎬	Equator pafs throug the Me-	⎨ 2. 54 ⎬	the middle of
wars till the	⎬ 3 ⎬	ridian, I find the Colure cut	⎬ 20.52 ⎬	the *Gnomon.*
Index points to	⌊ 4 ⌋	the *Gnomon, Semi-Circle* in	⌊ 64.36 ⌋	*Semi-Circle.*

   And thefe are the diftances of the After-noon Hour-lines; and muft all but the 1 a clock Hour-line be prickt down at their re- fpective diftances on the Eaft fide the Plane, *viz.* from B to D : But becaufe the Colure comes not to the middle of the *Gnomoni- cal Semi-Circle* before the firft 15. degrees of the Equator pafs through the Meridian after 12. therefore the 1 a clock muft ftand 5. degrees 45. minutes on the Weft fide of the Plane : And for this caufe I made diftinction with a line between the 1 a clock

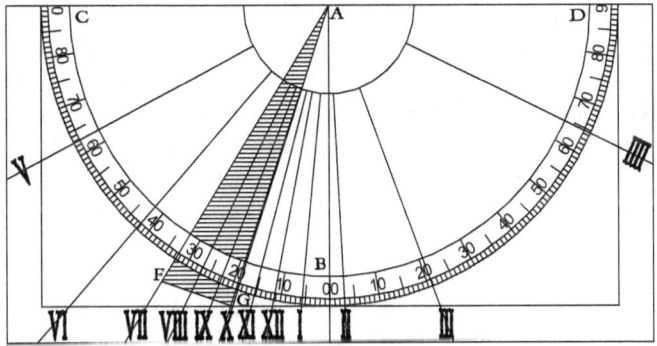

and the 2 a clock, in the foregoing Table. Then I draw lines from the Center A through every one of thefe pricks in the Semi-Circle, and they fhall be the Hour lines of this Declining Reclining Plane.

   Having drawn the Hour-lines, I remove the Quadrant of Al- titude to the degree of the Horizon oppofite to the degree it was at before, *viz.* to 30. degrees from the South Weftwards, which is fo much as the Plane declines Eaftwards; But I let the *Gno- monical Semi Circle* ftand as it did : And turning about the
Globe

Globe till the Colure cut the Quadrant of Altitude in 20. degrees counted from the Horizon upwards, *viz.* the degrees of Reclination, I find 18. degrees 40. minutes contained between the
middle of the *Gnomonical Semi Circle* and the *Braſen Meridian,*
which is the diſtance of the *Subſtyle* from the Perpendicular ;
And I find the *Gnomonical Semi Circle* cut the Colure in 13.
degrees 49. minutes from the Pole, which is the Height that the
Style muſt be raiſed over the *Subſtyle* ; Therefore I prick off in
the Semi Circle on the Plane, the diſtance of the *Subſtyle* 18. degrees 40. minutes from the Perpendicular Weſtwards, becauſe
this Plane declines Eaſtwards : And from the Center A, I draw
through that prick the line A E, which ſhall be the *Subſtyle,* and
from this *Subſtyle* (either way) I count in the Semi Circle on
the Plane 13. degrees 49. minutes, and there make a Prick :
Then from the Center A, I draw through that Prick the line A
F, to repreſent the *Style* or *Gnomon* : Then I let fall the Per-
pendiculer F G upon the *Subſtyle* A G ; So is a Triangle made ;
which if it be erected Perpendicularly upon the *Subſtyle* A G,
the *Style* A F ſhall be Parallel to the Axis of the World, and
caſt a ſhadow upon the Hour of the Day.

    Having made this Dyal, you have made four ſeveral Dyals,
whereof this is one : And his oppoſite, *viz.* North Declining
Weſtwards 30. degrees Inclining to the Horizon 70. degrees is
another. The South Declining Weſtwards 30. degrees Reclining from the *Zenith* 20. degrees is another : And his oppoſite,
viz. North Declining Eaſtwards 30. degrees Inclining to the
Horizon 70. degrees is the other.

PROB. XIII.

To make a Dyal *upon a* Declining Inclining Plane.

THe Precepts for making theſe Dyals are delivered in the
foregoing Probleme : Therefore we ſhall at firſt come
to an Example.

    I would make a Dyal upon a Plane in *Londons* La-
titude Declining from the South Weſtwards 25. degrees and Inclining towards the Horizon by the ſpace of an Arch containing
14. degrees. Having firſt diſcribed on the Plane a Semi Circle, as was directed Prob. 4. I rectifie the Globe, Quadrant of
Altitude,

Altitude, Colure, and Hour Index, as by the fame Probleme, and bring the lower end of the Quadrant of Altitude to the degree of the Planes Declination, *viz.* to 25. degrees counted from the South Weftwards, and the ends of the *Gnomonical Semi Circle* to the degree of *Azimuth* the Plane lies in, *viz.* to 25. degrees from the Weft Northwards, and the middle of the *Gnomonical Semi Circle* to the degree of the Planes Inclination, *viz.* 14. degrees counted from the *Zenith* downwards on the Quadrant of Altitude. Then counting the degrees of the *Gnomonical Semi Circle* contained between the middle of the fame and the *Brafen Meridian,* I find 5. degrees 30. minutes : Thefe 5. degrees 30. minutes fhews the diftance of the 12 a clock line from the Perpendicular; Therefore I number in the Semi Circle difcribed on the Plane, from the Perpendicular Weftwards, (Becaufe the middle of the *Gnomonical Semi Circle* lies Weftwards on the Globe) from the Meridian. And for finding all the Fore-Noon Hour-diftances

I turn the ⎧11⎫ a clock, or till 15 degr. of the ⎧20. 5⎫ counted from
Globe Eaft- ⎨10⎬ Equa. pafs throug the Meri- ⎨36. 57⎬ the middle of
wards till the ⎨ 9 ⎬ dian, and find the Colure cut ⎨56. 24⎬ the *Gnomon.*
Index points to ⎩ 8 ⎭ the *Gnomon. Semi-Circle* in ⎩76. 31⎭ *Semi-Circle.*

And thefe are the diftances of all the Fore Noon Hour lines ; to which feveral diftances I make pricks on the Weft fide the Semi Circle on the Plane, *viz.* from B to C.

The After Noon Hour lines are found by bringing the Colure again to the Meridian, and the Index of the Hour Circle to 12. For then

turning the ⎧1⎫ a clock, or till 15. ⎧ 6.20⎫ counted from
Globe Weft- ⎨2⎬ degrees of the Equa- ⎨18. 2⎬ the middle of
wards till ⎨3⎬ tor pafs through the ⎨28. 45⎬ the *Gnomoni-*
the Index ⎨4⎬ Meridian, I find the ⎨39. 56⎬ *cal Semi Cir-*
points to ⎨5⎬ Colure cut the *Gno-* ⎨52. 30⎬ *cle.*
⎨6⎬ *monical Semi-Circle* ⎨67. 19⎬
⎩7⎭ in ⎩84. 13⎭

And thefe are the diftances of the After Noon Hour lines, which I alfo prick down at their refpective diftances from the Perpendicular Eaftwards, *viz.* from B towards D on the Plane; and by drawing lines from the Center A through all the Pricks, I have all the Hour lines that this Plane will admit of.

Z

Having

Having drawn the Hour lines, I remove the lower end of the Quadrant of Altitude to the degree of the Horizon oppoſite to the degree it was at before, *viz.* to 25. degrees from the North Eaſtwards, which is ſo much as the Declination is Weſtwards; but I let the *Gnomonical Semi Circle* ſtand as it did, and turn about the Globe till the Colure cut the Quadrant of Altitude in 14. degrees counted from the Horizon upwards, which is the Inclination of the Plane : Then I find 24. degrees 3. minutes comprehended between the middle of the *Gnomonical Semi Circle* and the *Braſen Meridian,* which is the diſtance of the *Subſtyle* from the Perpendicular: and this diſtance I count Weſtwards on the Plane becauſe the middle of the Semi Circle lies Weſtwards on the Globe and draw the line A G through it for the *Subſtyle* : And I find the *Gnomonical Semi Circle* cut the Colure in 48. degrees 5. minutes, for the Heigth that the *Style* muſt be Elevated over the *Subſtyle* : Therefore I make a prick on the Plane 48. degrees 5 minutes diſtant from the *Subſtyle,* and through that prick I draw the line A F to repreſent the *Style* or *Gnomon* ; Then I let fall the Perpendicula F G upon the *Subſtyle* A G, ſo is there a T riangle made; which if it be erected

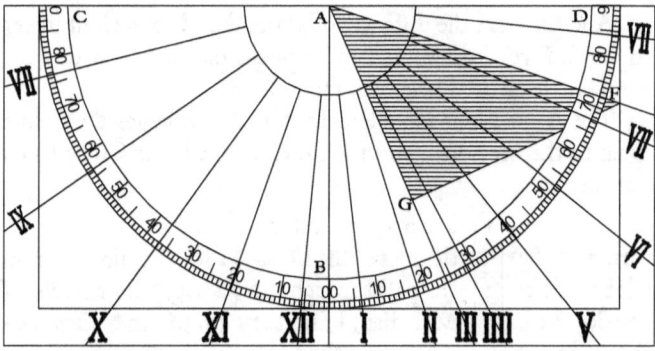

Perpendicularly upon the *Subſtyle* A G, the *Style* A F ſhall be Parallel to the Axis of the World, and caſt a ſhadow upon the Hour of the Day.

Having made this Dyal you have alſo four Dyals made ; as well as in the former Probleme : For this is one and its oppoſite *viz.* North declining Eaſtwards 25. degrees Reclining 76. degrees is another ; The South declining Eaſtwards 25. degrees

Indi-

Inclining 14. degrees is another; and its oppofite, *viz.* North declining Weftwards 25. degrees Reclining 76. degrees is another.

PROB. XIV.

To find in what Place of the Earth *any manner of Plane that in your* Habitation *is not* Horizontal, *fhall be* Horizontal.

IT was faid in the *Preface* that all manner of Planes however fcituate are Parallel to fome Country or other on the Earth: Therefore all manner of Planes are indeed Horizontal Planes: and the diftances of the Hour lines to be infcribed on them may be found as the diftances of the Hour lines of the Horizontal Dyal in Prob. 3. It refts now to learn in what place of the Earth any Plane that is not Horizontal in your Habitation fhall become Horizontal: And for help of your underftanding herein, Take thefe following Rules.

1. If your Plane be Erect Direct North, or South, it fhall be an Horizontal in the fame Longitude at 90. degrees diftance on the Meridian, (counted from the *Zenith* of your Place,) through the Equinoctial. See an Example of this in Prob. 3. where I have reduced an Erect Dyrect Dyal to an Horizontal. Thus an Erect Plane under the Pole is an Horizontal under the Equator; and an Erect Direct in 80. degrees North Latitude is in the fame Longitude an Horizontal at 10. degrees South Latitude: An Erect Direct in 70. degrees North Latitude, is in the fame Longitude, an Horizontal at 20. degrees South Latitude: and fo to any other degrees of Latitude (as aforefaid) till you come to 45. degrees Latitude where an Erect is an Horizontal, and an Horizontal an Erect. Only as the Hour lines of the Horizontal (being turned downwards) are numbred from the right hand towards the left, in the Erect Direct Dyal they are numbred from the left hand towards the Right.

2. If your Plane be Erect Declining, it fhall be an Horizontal Plane at that point on the Globe which is againft the degree of Declination, found in the Horizon.

But note, If your Plane declines Weftwards, the Sun comes fooner to the Meridian of it, then to the Meridian of the Place

where

where it becomes an Horizontal Plane ; and that by ſo many Hours or minutes as the degrees of the difference of Longitude between the two Places converted into Time amounts to. If it declines Eaſtwards, the Sun comes ſo much later to the Meridian of it : And for this Cauſe (though the making this Dyal be the ſame with an Horizontal Dyal for another Place, yet in Reſpect of Time) there will be a difference between them.

Example.

I would make the South Dyal Declining Eaſt 27. degrees, as in Prob. 9. by the Plane of the Horizon : Firſt I ſeek in what Place of the Earth it ſhall become an Horizontal Plane : Thus, I Elevate the Pole of the Globe 51 $\frac{1}{2}$ degrees above the Horizon, and bring the Vernal Colure to the Meridian, then I count from the South point in the Horizon Eaſtwards 27. degrees, and on the point on the Globe directly againſt thoſe 27. degrees I make a prick for the Place where a Plane that declines 27. degrees from the South Eaſtwards at *London* ſhall be Horizontal; or which is all one, this Declining Plane at *London* ſhall ly in the Horizon of that Prick : This Prick for diſtinction ſake we ſhall hereafter call the Horizontal Place: Then by Prob. 1. of the Second Book, I examine the Latitude and Longitude of this Horizontal Place, and find Latitude 33. 40. South; and Longitude from the Colure 33. degrees, which is the difference of Longitude between *London* and the Horizontal Place : which being converted into Time by allowing for every 15. degrees 1. hour of Time, gives 2 hours 12. minutes that the Sun comes ſooner to the Meridian of the Horizontal Place, then to the Meridian of the Plane at *London:* ſo that when it is 12 a clock there, it will be but 9. a clock 48. minutes here; when 12 a clock here, it will be 2 a clock 12. minutes There, &c.

Having thus found in what Longitude from *London* and Latitude this Plane is Parallel to the Horizon, I ſeek the diſtances of the Hour-lines upon the Planes of the Horizon Thus, I Elevate the Pole of the Globe to the Height of the Pole in the Horizontal Place, *viz.* 33. degrees 40. minutes, and bring the Horizontal Place on the Globe to the Meridian, and the Index of the Hour Circle to 12. Then I examine the degree of the Horizon the Colare cuts, and find it 19 $\frac{3}{4}$ from the South Weſtwards. This 19 $\frac{3}{4}$
degrees

degrees refprefents the Meridian line of the Horizontal Place : And alfo the *Subftylar line* here at *London;* Therefore this 19 ¾ degrees I count from the Perpendicular A B of the Plane, and from the Center A draw the line A G through them ; Becaufe from this line on the Plane all the Hour lines muft be numbred, and not (as all along hitherto) from the Perpendicular of the Plane, Then

turning the Globe Eaft-wards till the Index of the Hour Circle points to {11, 10, 9, 8, 7, 6, 5} a clock, or till 15. degr. of the Equator pafs through the Meridian, I find the Colure cut the Horizon in {10. 2, 0. 45, 6. 12, 15. 25, 26. 11, 40. 30, 60. 0} from the Meridian.

And thefe are the diftances of the Forenoon Hour lines: which diftances I tranffer by pricks to the Plane. But as in Prob. 9. I fought the diftances from the Perpendicular on the Plane, fo now in this Cafe (as was faid before) I feek them from the *Subftyle,* and through thefe pricks I draw lines from the Center, as in other Dyals, and thefe lines fhall be the Fore Noon Hour lines.

To find the Afternoon Hour diftances, I bring the Horizontal Place on the Globe again to the Meridian, and the Index of the Hour Circle to 12. and

turning the Globe Weftwards till the Index points to {1, 2, 3, 4} a clock, or till 15 degr. of the Equator pafs through the Meridian, I find the Colure cut the Horizon in {31. 5, 46. 32, 68. 5, 95. 37} counted from the Meridian.

And thefe are the diftances of all the Afternoon Hour lines; which I alfo transfer to the Plane, counting them from the *Subftyle,* and draw lines from the Center A through thefe diftances; and thefe lines fhall be all the Afternoon Hour lines.

Then from the *Subftyle* I count the degrees and minutes of the Latitude of the Horizontal Place, *viz.* 33. degrees 40. minutes, and through thefe degrees and minutes I draw the line A F from the Center A, for the *Style* : Then from the *Style* I let fall the Perpendicular F G upon the *Subftyle,* fo is there a Triangle made; which if it be erected Perpendicularly upon the *Subftyle* A G, the *Style* A F fhall be Parallel to the Axis of the World, and caft a fhadow upon the Hour of the Day.

Z 3   3. If

3. If your Plane be a Direct Recliner, Seek in the Longitude of your Place the Complement to 90. of your Planes Reclination: For there a Direct Recliner becomes an Horizontal Plane.

4. If your Plane be a Declining Recliner : The Globe and Quadrant of Altitude Rectified, Bring your Habitation on the *Terreftrial* Globe to the Meridian, and the Quadrant of Altitude to the Declination, as by the fecond Rule in this Probleme; and count upwards on the Quadrant of Altitude the Reclination, and there make a prick on the Globe by the fide of the Quadrant of Altitude, for at that prick on the Globe the Declining Recliner fhall become an Horizontal Plane. Then examine the Latitude of that prick as by Prob. 1. of the fecond Book, and the difference of Longitude, as by Prob. 9. of the third Book : And convert the difference of Longitude into Time, by allowing for every 15. degrees 1. hours Time, for every degree 4. minutes Time, and fo proportionably, fo fhall you know what Hours and Minutes the Sun comes fooner or later to the Meridian of your Habitation then to the Meridian of that Place where it becomes an Horizontal Plane: Sooner, if the Globe were turned Eaftwards; but Later if it were turned Weftwards.

Having thus found out where this Plane becomes Horizontal, make your Dyal to this Plane, as by the fecond Rule in this Probleme: Find alfo the *Style* as is there directed.

5. If your Plane be a Declining Incliner, The Globe and Quadrant of Altitude Rectified, Bring the Colure to the Meridian, and the Quadrant of Altitude to the degree of the Horizon oppofite to the degree of the Planes Declination, and count upwards on the Quadrant of Altitude the degrees of Inclination, and make a prick there ; For in the *Antipodes* of that prick (found as by Prob. 29. of the Second Book) that Declining Incliner fhall become an Horizontal Plane. Then find the Latitude and difference of Longitude of this *Antipodes,* by the former Rule, and make a Dyal to that Plain as by the fecond Rule in this Probleme. Find alfo the *Style* as therein is directed.

PROB.

PROB. XV.

To make a Dyal *on the* Ceeling *of a* Room, *where the* Direct Beams *of the* Sun *never come.*

Find fome convenient place in the Tranfum of a Window to place a fmal round peece of Looking-Glafs, about the bignefs of a Groat, or lefs; fo as it may ly exactly Horizontal : The point in the middle of this Glafs we will marke A, and for diftinctions fake (with Mr *Palmer*) call it *Nodus:* Through this *Nodus* you muft draw a Meridian line on the Floor, Thus: Hang a Plumb line in the Window exactly over *Nodus,* and the fhadow that that Plumb line cafts on the Floor juft at Noon will be a Meridian line ; Or you may find a Meridian line otherwife, as by the *Preface.* Having drawn the Meridian line on the Floor : find a Meridian line on the Ceeling, thus: Hold a Plumb line to the Ceeling, over that end of the Meridian line next the Window; If the Plumbet hang not exactly on the Meridian line on the Floor, remove your hand on the Ceeling one way or other, as you fee caufe till it do hang quietly juft over it : and at the point where the Plumb line touches the Ceeling make a mark, as at B; that mark B fhall be directly over the Meridian line on the Floor: then remove your Plumb line to the other end of the Meridian line on the Floor, and find a point on the Ceeling directly over it, as you did the former point as at C, and through thofe two points B and C on the Ceeling ftrain and ftrike a line blackt with Smal Cole or any other Culler (as *Carpenters* do,) and that line B C on the Ceeling fhall be a Meridian line, as well as that on the Floor : Then examine the Altitude of the Equinoctial, as by Prob. *6.* of the Second Book you did the Meridian Altitude of the Sun; and faften a ftring juft on the *Nodus,* and remove that ftring in the Meridian line on the Ceeling till it have the fame Elevation in a Quadrant that the Equinoctial hath in your Habitation; and through the point where the ftring touches the Meridian line in the Ceeling fhall a line be drawn at right Angles with the Meridian, to reprefent the Equinoctial line. Thus in our Latitude the Elevation of the Equator being 38 ½ degrees; I remove the ftring faftned to the *Nodus* forwards or backwards in the Meridian line of the Ceeling, till the Plumb line of a

Quadrant

Quadrant, when one of the ſides are applyed to the ſtring, falls upon 38½ degrees: and then I find it touch the Meridian line at D
in the Ceeling : therefore at D I make a mark and through this
mark ſtrike the line D E (as before I did the Meridian line) to
cut the Meridian line at Right Angles : This line ſhall be the
Equinoctial line.

Then I place the Center of the Semi-Circle of Poſition upon
Nodus, and under-prop it ſo that the flat ſide of it may ly Parallel
to the ſtring when it is ſtrained between the *Nodus* and the E-
quinoctial, and alſo ſo as the ſtring may ly on the diviſion of the
Semi-Circle marked o, when it is held up to the Meridian line in
the Ceeling : Then removing the ſtring the ſpace of 15. degrees
in the Circle of Poſition to the Eaſtwards, and extending it to the
Equator on the Ceeling, where the ſtring touches the Equator
there ſhall be a point through which the 1 a clock Hour-line
ſhall be drawn : and Removing the ſtring yet 15. degrees further
to the Eaſtwards in the Semi-Circle of Poſition, and extending it
alſo to the Equator, where it touches the Equator there ſhall be
a point through which the 2 a clock Hour-line ſhall be drawn :
Removing the ſtring yet 15. degrees further to the Eaſtwards in
the Semi-Circle of Poſition, and extending it to the Equator,
there ſhall be a point through which the 3 a clock Hour-line
ſhall be drawn : The like for all the other After-Noon Hour
lines; ſo oft as the ſtring is removed through 15. degrees on the
Semi-Circle of Poſition, ſo oft ſhall it point out the After-Noon
diſtances in the Meridian line on the Ceeling.

The ſcituation of the Semi-Circle of Poſition cannot conveniently be ſhewn in this Figure, unleſs it be drawn by the Rules of
Perſpective ; Neither if it were would it ſuit with the other de-
monſtrations, except they were drawn by the ſame Rules alſo;
which to do would be hard for young Learners to underſtand :
Therefore I have left out the Semi-Circle of Poſition in this Figure and refer you for a demonſtration thereof to the ſixth Probleme ; For even as the lines drawn through every 15. degrees
of the Semi-Circle there, denote in a Contingent line the diſtance of any Hour line from the Meridian line, even ſo a line
drawn through every 15. degrees of the Semi-Circle of Poſition
poſited (as aforeſaid) point out in the Equinoctial line on the
Ceeling the diſtance of each reſpective Hour line from the Meridian line.

Having

Having thus found out the points in the Equator through which the After-Noon Hour-lines are to be drawn, I may find the Fore-Noon Hour diftances alfo the fame way, *viz.* by bringing the ftring to the feveral 15. degrees on the Weft fide the Semi-Cirde of Pofition; or elfe I need only meafure the diftances of each Hour diftance found in the Equator from the Meridian line on the Ceeling; for the fame number of Hours from 12 have the fame diftance in the Equinoctial line on the other fide the Meridian both Before and Afternoon: The 11 a clock Hour diftance is the fame from the Meridian line with the 1 a clock diftance on the other fide the Meridian, the 10 a clock diftance the fame with the 2 a clock diftance, the 9 with the 3, &c. And thus the diftances of all the Hour lines are found out on the Equator.

Now if the Center of this Dyal lay within doores, you might draw lines from the Center through thefe pricks in the Equator, and thofe lines fhould be the Hour lines, as in other Dyals : But the Center of this Dyal lies without doores in the *Air*, and therefore not convenient for this purpofe : So that for drawing the

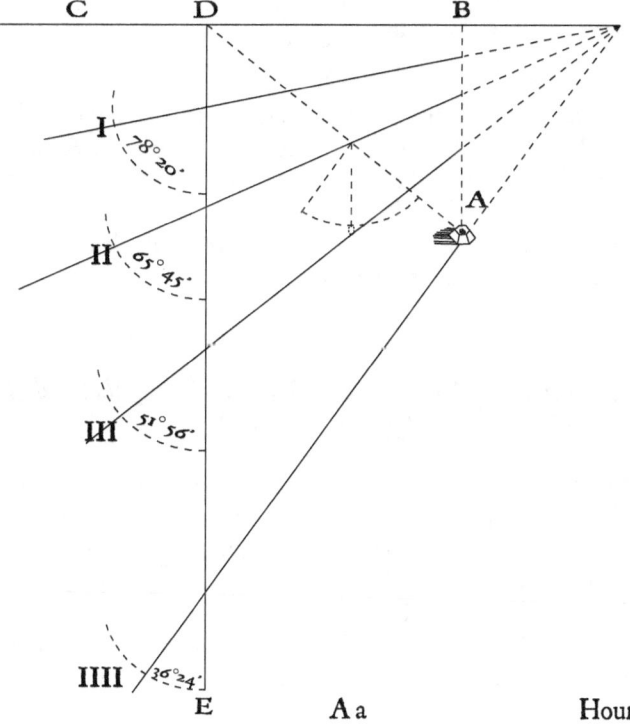

Hour lines you muſt conſider what angle every Hour line in an
Horizontal Dyal makes with the Meridian; that is, at what diſtance in degrees and minutes the Hour lines of an Horizontal
Dyal cut the Meridian; which you may examine as by Prob. 3. for
an Angle equal to the Complement of the ſame Angle muſt each
reſpective Hour-line make with the Equator on the Ceeling.

    Thus upon the point markt for each Hour diſtance in the E-
quinoctial line on the Ceeling, I diſcribe the Arches I, II, III,
IIII, as in the Figure, and finding the diſtance from the Meridian of the Hour-lines of an Horizontal Dyal to be according to
the third Probleme. Thus

$$\text{The} \begin{Bmatrix} 1 \\ 2 \\ 3 \\ 4 \end{Bmatrix} \text{a clock Hour-line} \begin{Bmatrix} 11.\ 40 \\ 24.\ 15 \\ 38.\ \ 4 \\ 53.\ 36 \end{Bmatrix} \begin{matrix} \textit{whoſe} \text{ Com-} \\ \text{plement} \quad \text{to} \\ 90. \text{ is} \end{matrix} \begin{Bmatrix} 78.\ 20 \\ 65.\ 45 \\ 51.\ 56 \\ 36.\ 24 \end{Bmatrix}$$

I meaſure in a Quadrant of the ſame Radius with thoſe arches
already drawn from the Equinoctial line

$$\text{for the} \begin{Bmatrix} 1 \\ 2 \\ 3 \\ 4 \end{Bmatrix} \text{a clock Hour} \begin{Bmatrix} 78.\ 20 \\ 65.\ 45 \\ 51.\ 56 \\ 36.\ 24 \end{Bmatrix}$$

and transfer theſe diſtances to the Arches drawn on the Ceeling:
For then ſtraight lines drawn through the mark in the Arch, and
through the mark in the Equator, and prolonged both waies to a
convenient length, ſhall be the ſeveral Hour-lines (aforeſaid;) And
when the Sun ſhines upon the Glaſs at *Nodus*, its Beams ſhall reflect upon the Hour of the Day.

PROB. XVI.

To make a Dyal *upon a ſolid* Ball, *or* Globe, *that ſhall ſhew
the* Hour *of the* Day *without a* Gnomon.

T He Equinoctial of this Globe, or (which is all one) the
    middle line muſt be divided into 24 equal parts, and mark-
ed with 1, 2, 3, 4 &c to 12. and then beginning again with 1, 2, 3,
&c. to 12. Then if you Elevate one of the Poles ſo many degrees
above an Horizontal line as the Pole of the World is Elevated
above the Horizon in your Habitation, and place one of the 12ˢ
directly to behold the North, and the other to behold the South:
                                          when

when the Sun fhines on it the Globe will be divided into two halfs, the one enlightened with the Sun-fhine, and the other fhadowed : and where the enlightned half is parted from the fhadowed half, there you fhall find in the Equinoctial the Hour of the Day; and that on two places on the Ball; becaufe the Equinoctial is cut in two oppofite points by the light of the Sun.

A Dyal of this fort was made by Mr *John Leek,* and fet up on a Compofite Columne at *Leaden Hall* Corner in *London,* in the *Majoralty* of Sr *John Dethick* Knight. The Figure whereof I have inferted, becaufe it is a pretty peece of Ingenuity, and may perhaps ftand fome Lover of Art in ftead, either for Imitation, or help of Invention.

PROB. XVII.

To make a Dyal *upon a* Glaſs Globe, *whoſe* Axis *ſhall caſt a ſhadow upon the* Hour *of the* Day.

FIrſt divide the Equinoctial of your Globe into 24 equal parts; and having a Semi-Circle cut out of ſome Braſs plate, or thin Wood to the ſame Diameter your Globe is of, or a very little wider : Apply this Semi-Circle to the Globe, ſo as the upper edge of each end of the Semi Circle may touch the Poles of the Globe, and the middle of the Semi Circle may at the ſame edge cut through ſome diviſion made in the Equinoctial: for then a line drawn by the edge of the Semi Circle thus poſited ſhall be a Meridian line ; The ſame way you muſt draw Meridian lines through every diviſion of the Equinoctial, and ſet figures to them, beginning with 1, 2, 3, 4, &c. to 12. and then beginning again with 1, 2, 3, 4, &c. to 12. again. This Globe being made of Glaſs, and having an Axis of Wyer paſſing through it from Pole to Pole, will be an Horizontal Dyal all the World over; if its Axis be ſet Parallel to the Axis of the World in the ſame Place; and one of the Meridians marked 12 be ſet ſo as it may directly behold the North point in Heaven, and the other the South point in Heaven; for then the Axis of the Globe ſhall caſt a ſhadow upon the Hour of the Day.

And if you divide the upper half of the Glaſs Globe from the under half, when the Axis ſtands Parallel to the Axis of the World, by a Circle drawn round about the Globe, that Circle ſhall repreſent the Horizon; and the Meridian lines drawn on the Globe ſhall be the Hour lines, and have in the Horizontal Circle the ſame diſtance from the 12 a clock line that the ſame reſpective Hour line was found to have, as by Prob. 3. of this Book.

But becauſe the ſhadow of this Axis will not be diſcerned through the Glaſs Body; therefore you may with Water and white Lead ground together, lay a Ground on the Inſide of the under half of the Glaſs to the Horizontal Circle (as Lookingglaſs makers do their Looking Glaſſes with *Tinfoil*) for then the ſhadow will appear.

Such a Glaſs Globe Dyal hath the Lord *Robert Titchborn* ſtanding in his *Garden* ſupported by *Atlas*.

The End of the Fifth Book.

The Sixth BOOK.

Shewing the Practical Use of the

GLOBES:

Applying them to the Solution of

Spherical Triangles.

P R Æ F A C E.

THe Solution of Spherical Triangles *is to know the length of its Sides, and the width of its Angles.* *These have already by many learned Men been taught, to be performed by a* Canon of Sines and Tangents; *and also by many Instruments some serving as Tables of* Sines and Tangents, *such as are the* Sectors, Scales, the Spiral line, &c. *and others serving to represent the* Globe; *such as be the* Mathematical Jewel, Astrolabium Catholicum, *and several other Projections of the* Sphear. *But none hath as yet taught the Solution of* Spherical Triangles *by the* Globe *it self; though it be the most natural, and most demonstrative way of all, and indeed ought first to be learnt before the Learner enters upon any other way.*

To

To this *Authors of* Trigonometry *agree, for the most of them in their Books give Caution that the Learner be already sufficiently grounded in the Principles of the* Globe: *For those* Lines *or* Circles *which either in* Tables *or other Instuments your force your Imagination to conceive represents your* Line *or* Circle *in question, those* Lines *and* Circles *I say, you have Actually and Naturally discribed on the* Globe, *and therefore may at a single Operation, or perhaps only by a sudden inspection, have an Answer annexed, according as the nature of your Question shall require : and that more Copiously then by* Tables *of* Sines *and* Tangents : *For therein you find but one Question at once resolved : but by the* Globe *you have alwaies two resolved together.*

Of the Parts and Kindes of Spherical Triangles.

THEOREMS.

1. ALL Spherical Triangles are made of six parts; Three Sides, and three Angles. The Sides are joyned together at the Angles, and measured by degrees of a Great Circle, from one end to the other. The Angles are the distance of the two joyned sides : and they are also measured by an Arch of a Circle, discribed on the Angular point. If any three of these parts be known, the rest may be found.

2. All Spherical Triangles are either Right Angled, or Oblique Angled. A Right Angle contains 90. degrees : An Oblique Angle either more, or less.

3. If a Spherical Triangle have one or more Right Angles; it is called a Right Angled Spherical Triangle. But if it have no Right Angle; it is called an Oblique Angled Spherical Triangle.

4. If an Oblique Spherical Triangle have one Angle greater then a Right Angle: it is called an Obtuse Angled Spherical Triangle : But if it have no Angle greater, it is called an Accute Angled Spherical Triangle.

5. In Right Angled Triangles the sides including the Right Angle are called Legs : And the side opposite to the Right Angle

gle is called Hypothenuſa. Thus the ſides
A B and A C in the following Triangle are
called Legs; and the ſide B C is called Hypothenuſa.

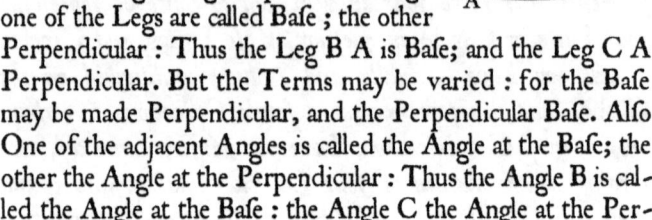

    6. In a Right Angled Spherical Triangle
one of the Legs are called Baſe ; the other
Perpendicular : Thus the Leg B A is Baſe; and the Leg C A
Perpendicular. But the Terms may be varied : for the Baſe
may be made Perpendicular, and the Perpendicular Baſe. Alſo
One of the adjacent Angles is called the Angle at the Baſe; the
other the Angle at the Perpendicular : Thus the Angle B is called the Angle at the Baſe : the Angle C the Angle at the Perpendicular.

PROB. I.

The Legs *of a* Right Angled Spherical Triangle *given; to
find the* Hypothenuſa, *and the two other* Angles.

T He Baſe of a Right Angled Spherical Triangle ſhall
in this following Treatiſe be alwayes placed on a Meridian, the Perpendicular on the Equator, the Hypothenuſa on the Quadrant of Altitude, and the Angle at the Baſe ſhall be meaſured in an Arch of the Horizon.

    Elevate the Equinoctial into the *Zenith,* ſo ſhall the Poles of
the Globe ly in the North and South points of the Horizon.

    Then count from the Equinoctial on the firſt Meridian, if you
uſe the *Terreſtrial Globe;* or on the Vernal Colure, if you uſe
the *Celeſtial,* becauſe they are divided from the Equinoctial
either way into 90. degrees ; and becauſe from thence the degrees of the Equinoctial are begun to be numbred : Count (I
ſay) from the Equinoctial the number of degrees the Baſe contains, and there make a prick : Then count in the Equinoctial
from the firſt Meridian the number of degrees the Perpendicular
contains, and make there a ſecond Prick : Bring that ſecond
Prick to the Braſen Meridian, ſo ſhall the firſt Meridian be ſepa-
rated from the Braſen Meridian by the quantity of an Arch e-
qual to the meaſure of the Perpendicular : Then having the
Quadrant of Altitude ſcrewed in the *Zenith,* turn it about till the
ſide of it cut the Prick made in the firſt Meridian ; ſo ſhall the
Triangle

Triangle be reprefented on the Globe. The Bafe fhall ly on the firft Meridian between the Equinoctial and the Quadrant of Altitude, the Perpendicular in the Equinoctial between the firft Meridian and the Brafen Meridian; and the Hypothenufa on the Quadrant of Altitude between the *Zenith* and the firft Meridian: and the number of degrees between each of thefe refpective Arches fhall be the meafure of each refpective Side. For the Angles ; The Right angle is known to be 90. degrees, by the fecond Theorem in the Preface. The meafure of the Angle at the Perpendicular is numbred between the Eaft point in the Horizon and the graduated edge of the Quadrant of Altitude : But to find the Angle at the Bafe you muft turn the Triangle, making the Perpendicular Bafe, and the Bafe Perpendicular.

Example.

Having the two Legs given A B 79. degreee 15. minutes, and C A 23. degrees 8 minutes, I would find the meafure of the Hypothenufa C B, and the Angles B C.

The Equinoctial Elevated, as before, I make A B Bafe, and C A Perpendicular, counting in the firft Meridian from the Equinoctial 79. degrees 15. minutes, and there I make a prick: Then I number in the Equinoctial from the firft Meridian 23. degrees 8. minutes, the length of the Perpendicular, and there I make a fecond Prick : This prick I bring to the Brafen Meridian, fo is the firft Meridian feparated
from the Bra-
fen Meridian
fo many de-
grees and mi-
nutes as is the

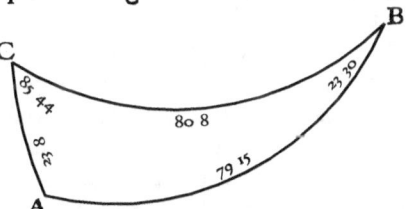

length of the Perpendicular C A; Then I fcrew the Quadrant of Altitude to the *Zenith, viz.* directly over the Equinoctial, and move it about till the edge of it touch the Prick made in the firft Meridian: So is the Triangle made on the Globe : And the number of degrees and minutes of the Quadrant of Altitude comprehended between the firft Meridian and the *Zenith* is the meafure of the Hypothenufa C B; which in this Example is 80. degrees 8. minutes : The number of degrees in the Horizon compre-

hended

hended between the Equinoctial and the Quadrant of Altitude is the meaſure of the Angle C, 85. degrees 44. minutes: the Angle A is a Right Angle, 90. degrees: And to find the Angle B, turn the Triangle, (all but the Letters;) Thus: As before A C was Baſe, ſo now I make B A Baſe; and as before A B was Perpendicular, ſo now C A ſhall be Perpendicular : ſo is your Triangle turned.

Now, as before I counted 79. degrees 15. minutes from the Equinoctial on the firſt Meridian, which was the length of that Baſe, ſo now I count 23. degrees 8. minutes on the firſt Meridian, which is the length of this Baſe, and there (as before) I make a Prick : and as before I counted 23. degrees 8. minutes on the Equinoctial from the firſt Meridian, which was the length of that Perpendicular ; ſo now I count 79. degrees 15. minutes on the Equinoctial, which is the length of this Perpendicular ; and there I make a prick on the Equinoctial: Then I bring this Prick (as before) to the Braſen Meridian, ſo ſhall the firſt Meridian be diſtant (as before) from the Braſen Meridian 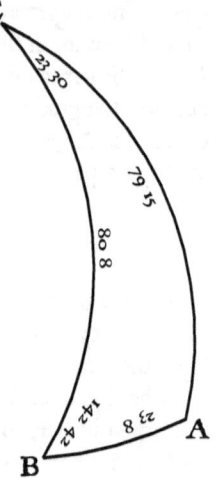 ſo many degrees and minutes as is the length of this Perpendicular, *viz.* 79. degrees 15. minutes: Then Having the Quadrant of Altitude ſcrewed to the *Zenith*, I turn it about till the edge of it touch the Prick made in the firſt Meridian at 23. degrees 8. minutes diſtant from the Equinoctial; ſo is the Triangle Turned : And ſo ſhall the Arch of the Horizon comprehended between the Equinoctial and the Quadrant of Altitude be the meaſure of the Angle C in the former Triangle, (but now made B) 23. degrees 30. minutes: you alſo ſee again the meaſure of the Hypothenuſa B C 80. degrees 8. minutes on the Quadrant of Altitude, counted between the *Zenith* and the firſt Meridian.

PROB. II.

A Leg *and the* Hypothenuſa *given, to find the* Reſt.

Example. The Leg given ſhall be C A in the former Triangle 23. degrees 8. minutes, The Hypothenuſa C B 80. degrees 8. minutes. The Equinoctial and Quadrant of Altitude Rectified, as by the laſt Probleme ; Number the Leg C A 23. degrees 8. minutes on the Equinoctial from the firſt Meridian, and there make a prick ; Bring this Prick to the Braſen Meridian ; Then number on the Quadrant of Altitude the Hypothenuſa B C 80. degrees 8. minutes from the *Zenith* towards the Horizon, and make there on the edge of the Quadrant of Altitude another prick : Then turn the Quadrant of Altitude about till the prick made on the edge of it touch the firſt Meridian; ſo ſhall the Triangle be made : The arch of the Equinoctial comprehended between the firſt Meridian and the Braſen Meridian, ſhall repreſent A C the Perpendicular ; the arch of the Quadrant of Altitude comprehended between the *Zenith* and the Firſt Meridian, ſhall repreſent B C the Hypothenuſa ; and the arch of the firſt Meridian comprehended between the Equinoctial and the Quadrant of Altitude ſhall repreſent B A the Baſe ; which was one Leg ſought, and is (as you will find) 79. degrees 15. minutes: The Angle C you will find in the Horizon 85. degrees 44. minutes: The angle A is the Right Angle 90. degrees : And to find the Angle B you muſt turn the Triangle, as you were directed in the former Probleme.

PROB. III.

The Hypothenuſa *and an* Angle *given, to find the* Reſt.

The Hypothenuſa given ſhall be B C of the Triangle in Prob. 1. 80. degrees 8. minutes, The Angle given ſhall be C 85. degrees 44 minutes : The Globe and Quadrant of Altitude Rectified, as by Prob. 1. Count the given Angle 85. degrees 44. minutes on the Horizon from the Equinoctial, and there place the Quadrant of Altitude : Then turn about the Globe till the firſt Meridian touch 80. degrees 8.

minutes

minutes of the Quadrant of Altitude counted from the *Zenith* downwards, ſo ſhall the Triangle be made on the Globe : The Arch of the Equator comprehended between the firſt Meridian and the Braſen Meridian ſhall ſhew the length of the Perpendicular C A 23. degrees 8. minutes; the Arch of the firſt Meridian comprehended between the Equinoctial and the Quadrant of Altitude ſhall ſhew the length of the Baſe A B 79. degrees 15 minutes ; the Right Angle made at the Interſection of the Braſen Meridian with the Equinoctial is 90. degrees : and to find the meaſure of the Angle B you muſt turn the Triangle, as you were directed Prob. 1.

PROB. IIII.

A Leg *and* Angle *adjoyning given, to find the* Reſt.

IN the Triangle of Prob. 1. The Leg given ſhall be C A 23. degrees 8. minutes, the Angle adjoyning ſhall be C 85. degrees 44 minutes : The Globe and Quadrant of Altitude Rectified, as by Prob. 1. I turn about the Globe till the firſt Meridian be diſtant from the Braſen Meridian 23. degrees 8. minutes, the length of the Leg C A : Then I count in the Horizon from the Equinoctial 85. degrees 44. minutes, the meaſure of the Angle C ; ſo is the Triangle made on the Globe. The Arch of the firſt Meridian comprehended between the Quadrant of Altitude and the Equinoctial ſhall ſhew the length of the Baſe A B 79. degrees 15. minutes ; The Arch of the Quadrant of Altitude comprehended between the *Zenith* and the firſt Meridian ſhall ſhew the length of the Hypothenuſa C B 80. degrees 8. minutes ; The Right Angle made at the Interſection of the Equinoctial and the Braſen Meridian is 90. degrees : And to find the meaſure of the Angle B, you muſt turn the Triangle, as you were directed Prob. 1.

PROB. V.

A Leg *and the* Angle *oppoſite given, to find the* Reſt.

IN the Triangle of Prob. 1. the Leg given ſhall be A B 79. degrees 15. minutes, the Angle oppoſite ſhall be C 85. degrees
44. mi-

44. minutes. The Globe and Quadrant of Altitude Rectified, as by Prob. 1. I bring the Quadrant of Altitude to 85. degrees 15. minutes of the Horizon, the meafure of the Angle C :
Then I turn the Globe till 79. degrees 15. minutes of the firft
Meridian (which is the meafure of the Leg A B) touch the Quadrant of Altitude, fo is the Triangle made on the Globe. The
Arch of the Equinoctial comprehended between the firft Meridian
and the Brafen Meridian fhews the length of the Leg C A 23. degrees 8. minutes; the Arch of the Quadrant of Altitude com-
prehended between the *Zenith* and the firft Meridian, fhall fhew
the length of the Hypothenufa C B 80. degrees 8. minutes: The
Right Angle made at the Interfection of the Equinoctial and the
Brafen Meridian is 90. degrees : And to find the meafure of the
Angle B, you muft turn the Triangle, as you were directed in
Prob. 1 :

PROB. VI.

The Angle *given, to find the* Sides.

IN this Cafe you muft turn the Angles into Sides, making an
Oblique Triangle on the Globe, whofe Sides fhall be equal
to the given Angles : fo fhall the Angles of this Triangle
found, be the meafure of the Sides required.

Example.

In the Triangle of Prob. 1. The Angle A is 90. degrees, the
Angle B 23. degrees 30. minutes, the Angle C. 85. degrees 44.
minutes : The Globe and Quadrant of Altitude Rectified, as by
Prob. 1. I fet the Right Angle A 90. degrees on the Brafen Meridian, between the Pole and the Equinoctial; For the Angle B I
number downwards on the Quadrant of Altitude 23. degrees
30. minutes, which fhall be the fide reprefenting that Angle : for
the Angle C I number on the firft Meridian from the Pole towards the Equinoctial 85. degrees 44. minutes, which fhall be
the fide reprefenting that Angle : Then I turn the Globe and
Quadrant of Altitude till I can joyn the 23. degrees 30. minutes
counted before on the Quadrant of Altitude and this 85. degrees
44. minutes counted in the firft Meridian together; So is a Tri-

angle made on the Globe; whoſe ſides being equal to the Angles given, ſhall have its Angles equal to the ſides required : Thus the Arch of the Equinoctial comprehended between the firſt Meridian and the Braſen Meridian ſhall be found 23. degrees 8. minutes, the meaſure of the ſide A C : The Arch of the Horizon contained between the neareſt Pole and the Quadrant of Altitude ſhall be found 79. degrees 15. minutes, the meaſure of the ſide B A : And to find the Hypotenuſa B C, you have now *Data's* enough, either to find it as by ſome of the former Problemes ; or elſe you may find it by turning the Triangle as by Prob. 1.

    Theſe Caſes of Right Angled Spherical Triangles may be wrought otherwaies by the Globe, If you alter its Poſition ; making the North or South points of the Horizon, *Zenith* ; or elſe the Poles of the World, or the Poles of the *Ecliptick* ; and uſe the Circle of Poſition inſteed of the firſt Meridian or Circles of Longitude: But theſe Inſtructions together with a little Practiſe, are (I judge) ſufficient ; Therefore I ſhall refer Varieties to the Studies of the Induſtrious Studient.

Of Oblique Triangles.

PROB. VII.

The three Sides *given, to find the* Angles.

E Levate the Pole of the Globe above the Horizon to the Complement of one of the given Sides and ſcrew the Quadrant of Altitude in the *Zenith*, ſo ſhall that given Side be comprehended between the Pole and the Quadrant of Altitude; Then count from the Pole upon the firſt Meridian the meaſure of the Second Side, and there make a prick ; Count alſo from the *Zenith* upon the Quadrant of Altitude downwards the meaſure of the third Side, and make there on the edge of the Quadrant of Altitude another prick ; Then turn the Globe and Quadrant of Altitude till you can joyn theſe two pricks together; ſo ſhall your Triangle be made on the Globe : And then the number of degrees of the Equinoctial comprehended between the firſt Meridi-

an

an and the Brafen Meridian fhall be the meafure of the Angle
at the Pole : The Arch of the Horizon comprehended between
the Quadrant of Altitude and the interfection of the Brafen Meridian with the Horizon on that fide the Pole is elevated, fhall be
the meafure of the fecond Angle : And for finding the third Angle, you muft turn the Triangle, as by Prob. 1.

Example.

In the Triangle A B C annexed, The Side A B contains 38.
degrees 30. minutes, the fide B C 25. degrees, and the fide A C
60. degrees; I would meafure
thefe Angles ; I place one of
thefe fides upon the Meridian,
viz, A B 38. degrees 30 minutes, the Complement of 38.
degrees 30. minutes is 51 degrees 30. minutes; Therefore
I Elevate the Pole 51. degrees 30. minutes above the Horizon,

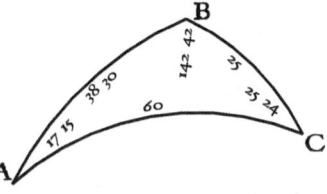

fo fhall the *Zenith* be diftant from the Pole 38. degrees 30. minutes; here I fcrew the Quadrant of Altitude and count downwards on it the meafure of the fide B C 25. degrees, and there I
make a prick : Then from the Pole I count on the firft Meridian 60. degrees, the meafure of the fide A C, and there I make a-
nother prick : Then I turn the Globe and Quadrant of Altitude backwards or forwards till thefe two pricks are joyned together ; fo fhall the Triangle A B C be made on the Globe :
The arch of the Brafen Meridian comprehended between the
Pole and *Zenith* fhall reprefent the fide A B; the Arch of the
Quadrant of Altitude comprehended between the firft Meridian
and the Brafen Meridian fhall reprefent the fide B C ; and the
Arch of the firft Meridian comprehended between the Pole and
the Quadrant of Altitude fhall reprefent the fide A C ; The Pole
fhall reprefent the Angle A, the *Zenith* the Angle B ; and the
interfection of the firft Meridian with the Quadrant of Altitude
fhall reprefent the Angle C. The Angle at the Pole is meafured
in the Equator; for the degrees comprehended between the firft
Meridian and the Brafen Meridian being 17. degrees 15. minutes
fhews 17. degrees 15. minutes to be the meafure of the Angle
A. The Angle at the *Zenith* is meafured in the Horizon ; for
the

the degrees comprehended between the Interſection of the Braſen Meridian with the Horizon on that ſide the Pole is Elevated being 142. degrees 42. minutes, ſhews that 142. degrees 42. minutes is the meaſure of the Angle B. Thus two angles are found; the third is wanting : which I find thus.

I turn the Triangle, placing either A or C in the *Zenith. Example:* I place A at the *Zenith,* which before was at the Pole ; ſo ſhall C be at the Pole, and B at the Interſection of the firſt Meridian and the Quadrant of Altitude, and the ſide A C ſhall be comprehended between the Pole and *Zenith :* The ſide A C contains 60. degrees; its Complement to 90 is 30. degrees ; therefore I Elevate the Pole of the Globe 30. degrees above the Horizon ; ſo ſhall 60. degrees be in the *Zenith;* therefore to 60. degrees I ſcrew the Quadrant of Altitude and count on it downwards the meaſure of the other ſide next the *Zenith, viz.* 38. degrees 30. minutes; and there I make a prick : Then from the Pole on the firſt Meridian I count the meaſure of the laſt ſide, *viz.* 25. degrees, and there I make another prick : Then I turn the Globe and Quadrant of Altitude (as before) till theſe two pricks joyn ; ſo is the Triangle altered on the Globe : For the Arch of the Braſen Meridian comprehended between the Pole and *Zenith* which before was 38. degrees 30. minutes, is now 60. degrees; the Arch of the Quadrant of Altitude Comprehended between the firſt Meridian and the Braſen Meridian, which before was 25 degrees, is now 38. degrees 30. minutes; and the Arch of the firſt Meridian comprehended between the Quadrant of Altitude and the Pole, which before was 60. degrees is now 25. degrees. Thus the Angle C being now at the Pole, its meaſure is found in the Equinoctial, *viz.* that Arch comprehended between the firſt Meridian and the Braſen Meridian, which is 25. degrees 24. minutes; and the meaſure of the Angle A, which is now in the *Zenith,* having its ſides, the one an Arch of the Braſen Meridian, the other an *Azimuth,* (or which is all one) an Arch of the Quadrant of Altitude, is meaſured in the Horizon, as all *Azimuths* are, and found 17. degrees 15. minutes, as before.

PROB.

PROB. VIII.

Two Sides *and the* Angle *contained between them given, to find the Reſt.*

EXample. In the former Triangle I have given the ſides A B, 38. degrees 30. minutes, A C, 60. degrees, and the Angle A 17. degrees 15. minutes.

The Method I have hitherto uſed is to place the given ſide up-on the Meridian between the Pole and *Zenith;* but becauſe the Angle at the Pole in this Example falls out ſo large that the Quadrant of Altitude will not reach the firſt Meridian; therefore I ſhall uſe another way to work this Probleme as well.

I Elevate the Pole of the Globe to the Com-plement of one of the given ſides ; ſuppoſe the ſide A B, which being 38. degrees 30. minutes, its Complement to 90. degrees is 51. degrees 30. minutes, ſo ſhall the *Zenith* be diſtant from the Pole 38. degrees 30. minutes, the meaſure of the ſide A B : The other ſide is 60. degrees, this 60. degrees I count from the Pole in the firſt Meridian, and there I make a prick: The Angle given is 17. degrees 15. minutes; this I count in the Equinoctial from the firſt Meridian, and this degree and minute in the Equinoctial I bring to the Braſen Meridian, ſo ſhall the firſt Meridian be ſe-parated from the Braſen Meridian 17. degrees 15. minutes: Then I ſcrew the Quadrant of Altitude to the *Zenith,* and bring the ſide of it to the prick made in the firſt Meridian: ſo ſhall the Tri-angle be made on the Globe. Then to find the unknown ſide B C, I count the number of degrees on the Quadrant of Altitude comprehended between the *Zenith* and the firſt Meridian, and find 25. degrees, which is the meaſure of the ſide B C : To find the meaſure of the Angle B, I count the number of degrees con-tained between the Interſection of the Meridian with the Horizon on that ſide the Pole is Elevated and the Quadrant of Altitude, and find 142. degrees 42. minutes, which is the meaſure of the Angle B : And to find the Angle C I turn the Triangle, as in Prob. 7.

C c PROB.

PROB. IX.

Two Sides *and an* Angle *oppoſite to one of them given, to find the* Reſt.

EXample. In the Triangle in Prob. 7. the Sides given are A B 38. degrees 30. minutes, and A C 60. degrees : The Angle given oppoſite to A C is B 142. degrees 42. minutes: I Elevate the Pole to the Complement of one of the given ſides; ſuppoſe A B, which being 38. degrees 30. minutes, its Complement to 90. degrees is 51. degrees 30. minutes; ſo is the *Zenith* diſtant from the Pole 38. degrees 30. minutes : To this 38. degrees 30. minutes I ſcrew the Quadrant of Altitude, and count in the Horizon from the Interſection of the Meridian with the Horizon on that ſide the Pole is Elevated the meaſure of the given Angle B, *viz.* 142 degrees 42. minutes, and to this number of degrees and minutes of the Horizon I bring the edge of the Quadrant of Altitude, then I count in the firſt Meridian from the Pole the meaſure of the ſide A C 60. degrees ; and there I make a prick, and turn about the Globe till that prick come to the edge of the Quadrant of Altitude, ſo is the Triangle made on the Globe. The degrees of the Quadrant of Altitude comprehended between the firſt Meridian and the *Zenith* being 25. degrees, is the meaſure of the ſide B C : The degrees of the Equinoctial comprehended between the firſt Meridian and the Braſen Meridian being 17. degrees 15. minutes, is the meaſure of the Angle A; and for finding the meaſure of the Angle C, I turn the Triangle, as in Prob. 7.

PROB. X.

Two Angles *and the* Side *comprehended between them given, to find the* Reſt.

EXample. In the Triangle of Prob. 7. the Angles given are A 17 degrees 15. minutes, and B 142. degrees 42. minutes, the ſide comprehended between them is A B 38 degrees 30. minutes; I Elevate the Pole to the Complement of the ſide A B which being 38. degrees 30. minutes, its Complement to 90 degrees

grees is 51. degrees 30. minutes, fo is the *Zenith* diftant from the Pole 38. degrees 30. minutes; to this 38. degrees 30. minutes I fcrew the Quadrant of Altitude, and count in the Horizon from the Interfection of the Meridian with the Horizon on that fide the Pole is Elevated the meafure of the given Angle B, *viz.* 142. degrees 42. minutes, and to this number of degrees and minutes of the Horizon I bring the edge of the Quadrant of Altitude; then I turn about the Globe till the firft Meridian is diftant from the Brafen Meridian 17. degrees 15. minutes of the Equinoctial, which is the meafure of the other given Angle ; So fhall the Triangle be made on the Globe: and the Arch of the Quadrant of Altitude comprehended between the firft Meridian and the *Zenith* fhall be the meafure of the fide B C 25. degrees, and the Arch of the firft Meridian comprehended between the Pole and its Interfection with the Quadrant of Altitude fhall be the meafure of the fide A C 60. degrees : The meafure of the Angle C is found by turning the Triangle, as in Prob. 7.

PROB. XI.

Two Angles *and a* Side *oppofite to one of them given, to find the* Reft.

EXample. In the Triangle of Prob. 7. the Angles given are A 17. degrees 15 minutes, and B 142. degrees 42. minutes, the fide given is B C 25. degrees, being the fide oppofite to the Angle A; the Angle A is made at the Pole of the Globe, and meafured in the Equator : Therefore I feparate the firft Meridian from the Brafen Meridian 17. degrees 15. minutes, fo doth the Pole reprefent the Angle A; the Angle B is made at the *Zenith*, and meafured in the Horizon ; therefore I count in the Horizon 142. degrees 42. minutes, and there I make a prick, to this prick I bring the edge of the lower end of the Quadrant of Altitude, (not minding to what degrees of the Meridian the upper end of it is placed) Then I count from the upper end of the Quadrant of Altitude 25. degrees downwards the meafure of th e fide B C and there I make a prick, and keeping the lower end of the Quadrant of Altitude to the prick made in the Horizon, I flide the upper end of it forwards or backwards till the prick on the Quadrant of Altitude come to the firft Meridian, fo fhall the Triangle be made

on the Globe : Then the Arch of the Brafen Meridian comprehended between the Pole and the upper end of the Quadrant of
Altitude fhall be the meafure of the fide A B 38. degrees 30. minutes; and the Arch of the firft Meridian comprehended between
the prick on the Quadrant of Altitude and the Pole fhall be the
meafure of the fide A C 60. degrees ; But the Angle C you muft
find by turning the Triangle ; as in Prob. 7.

In the working this Probleme I would have placed the given
fide B C 25. degrees upon the Brafen Meridian between the Pole
and *Zenith* ; but then the Angle B (being fo Obtufe) would
have had that fide which would be interfected by the Quadrant
of Altitude (*viz.* the firft Meridian under the Horizon, which
the Quadrant of Altitude cannot reach.

PROB. XII.

Three Angles *given, to find the* Sides.

THis Triangle is taught to be refolved by Mr *Palmer* on the
Planifphear ; Book 3. Chap. 19.

It is to be known (faith he) That if you go to the Poles of the
three great Circles whereof your Triangle is made, thefe Poles
fhall be the Angular points of a fecond Triangle ; and the two
leffer fides of this fecond Triangle fhall be equal to the two leffer
Angles of your firft Triangle : the greateft fide of the fecond
Triangle fhall be the fupplement of the greateft Angle of the firft
Triangle (that is, fhall have as many degrees and minutes as the
greateft Angle of the firft Triangle wanted of 180. degrees (fee
Pitifcus Trigonometry Lib. 1. *Prop 61.*

This fecond Triangle therefore (all whofe fides are known from
the Angles of the firft) you fhall refolve by Prob. 7. And having by that Probleme found the Angles of this fecond Trian-
,, gle, know that the two leffer Angles of the fecond Triangle
,, fhall be feveral and refpectively equal to the two leffer fides of
,, the firft Triangle, (and the leaft Angle to the leaft fide, the
,, middle Angle to the middle fide) and the greateft Angle of this
,, fecond Triangle being fubtracted out of 180. degrees, fhall
,, leave you the greateft fide of your firft Triangle.

Example. If the Angles be given 142. degrees 42. minutes,
17. degrees 15. minutes, and 25. degrees 24. minutes ; and the
fides

fides be enquired. Draw by aim a rude Scheam of this firft Triangle, writing in the Angle A 17 degrees 15. minutes, in B 142. degrees 42. minutes, in C 25. degrees 24. minutes ; fuppofing the fides yet unknown: then draw under this by aim alfo, a Scheam of the fecond Triangle, fetting his Bafe Parallel with the Bafe of the firft and making the Bafe of the fecond fhorter then the Bafe of the firft. Set alfo B at the Vertical Angle, and A C at the Bafe; as in the firft Triangle.   Then fay,

   Becaufe A in the firft Triangle is 17. degrees 15. minutes, therefore in the fecond Triangle B C (which fubtendeth A) fhall be 17. degrees 15. minutes: and becaufe C in the firft Triangle is 25. degrees 24. minutes, therefore in the fecond Triangle the fide A B (which fubtendeth C) fhall be 25. degrees 24. minutes; and becaufe B the greateft Angle in the firft Triangle, is 142. degrees 42. minutes, therefore in the fecond Triangle the fide A C (which fubtendeth B) fhall be the complement thereof to 180. degrees, *viz.* 37. degrees 18. minutes : Write now upon the fides of this fecond Triangle the quantities of the fides, fo is your fecond Triangle ready to be refolved, as by Prob. 7. Whereby you fhall find the Angles of the fecond Triangle, as I have expreffed them in the Scheam. A 25. degrees, C 38. degrees 30. minutes, B 120. degrees.

   Now laftly, I fay thefe Angles of the fecond Triangle thus found, give me the fides of the firft Triangle, which I feek, in this manner.

In the fecond Triangle.   In the firft Triangle.

$\begin{cases} \text{A is 25. degrees} \\ \text{C is 38. 30.} \\ \text{B is 120. 00.} \end{cases}$ Therefore $\begin{cases} \text{B C is 25. degrees} \\ \text{A B } \quad \text{38. 30.} \\ \text{A C 60. 00. Complement} \end{cases}$

of 120. degrees to 180. And thus by all the Angles given, we have found out all the fides, which was required.

   Having then the Angles of your firft Triangle given, and his fides now found; you fhall find his fcituation on the Globe thus: Place him as in Prob. 7. A B 38. degrees 30. minutes between

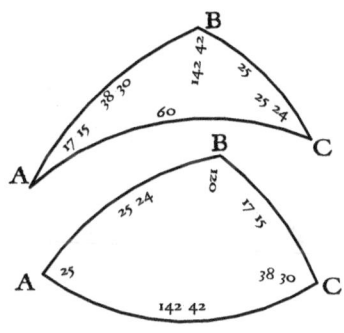

the Pole and *Zenith,* A C 60. degrees in the firft Meridian
feparated from the Brafen Meridian 17. degrees 15. minutes of
the Equinoctial, B C 25. degrees on the Quadrant of Altitude, counted from the *Zenith* when its lower end is applyed
to the 142. degrees 42. minutes of the Horizon: you fhall fay,
Becaufe the Eaft point of the Horizon is the Pole of the Arch A
B, therefore at the Eaft point of the Horizon fhall ftand the Angle C, which A B fubtendeth : Next follow the 142. degrees
42. minutes of *Azimuth* which maketh B C of your Triangle to the Horizon, and from thence number in the Horizon towards the Eaft point 37. degrees 18. minutes, the Complement of
the Angle A to 180. degrees, and number yet further 52. degrees
42. minutes beyond the Eaft point to make up 90, and there is the
Pole of the Arch B C : Therefore there fhall ftand the Angle
A, which B C fubtendeth. Then count in the Equator from the
firft Meridian 90. degrees, which will end under the Horizon, and
there make a prick ; for there is the Pole of the Arch or fide
A C. Therefore at that prick fhall ftand the Angle B, which A
C fubtendeth.

Here you fee your fecond Triangle made by the Poles of the
firft adjoyning to the Eaft point of the Globe: only the fide A B
is wanting: To get that, make a prick upon the Globe againft the
52. degres 42. minutes from the Eaft point of the Horizon found
before, to reprefent the Angle A: Then turn about the Globe and
Qudrant of Altitude till that prick and the prick made before
for the Angle B are both at once cut by the fide of the Quadrant of Altitude, and you will find 25. degrees 24. minutes of
the Quadrant of Altitude comprehended between the two
pricks, for the meafure of the fide A B.

PROB. XIII.

How to let fall a Perpendicular *that fhall divide any* Oblique Spherical Triangle *into two* Right Angled
Spherical Triangles.

This Probleme is much ufed when an Oblique Triangle
having two fides and an Angle given is to be folved by
the Cannon of *Sines* and *Tangents*: but by the Globe
it may be folved without it, as was fhewed Prob, 8, 9.

Yet

Yet becaufe letting fall a Perperdicular is fo frequent in all Authors that treat of *Trigonometry*, I have inferted this Probleme alfo.

In the Oblique Triangle
of the fromer Problemes
there is given the fides A B
38 $\frac{1}{2}$ degrees, and B C 25.
degrees, and the Angle C
25. degrees 24. minutes; It

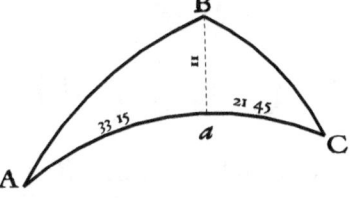

is required to let fall a Perpendicular as B *a* from the Angle B.
upon the Bafe A C; and to know both the meafure of this Perpendicular, and the parts it divides the Bafe into.

Therefore Elevate the Pole of the Globe above the Horizon
fo much as is the meafure of the Angle C, which in this Example
is 25. degrees 24. minutes, and bring the interfection of the firft
Meridian with the Equinoctial to the Eaft point of the Horizon ;
fo fhall the Angle at the Eaft point of the Horizon comprehended
between the Horizon and the firft Meridian be equal to the Angle C : then count in the firft Meridian from the Eaft point of
the Horizon the meafure of the fide B C 25. degrees, and having
the Quadrant of Altitude fcrewed to the *Zenith* bring the graduated edge of it to thefe 25. degrees, fo fhall the Arch of the Horizon comprehended between the Eaft point and the lower end of
the Quadrant of Altitude be the number of degrees that the Perpendicular falls upon the Bafe, counted from the Angle C to *a*,
which in this Example is 21 $\frac{3}{4}$ degrees, and the Arch of the Quadrant of Altitude comprehended between the Horizon and the
firft Meridian is the meafure of the Perpendicular B *a* 11. degrees.

And thus by letting fall this Perpendicular you have two Right
angled Spherical Triangles made, the one B *a* C, wherein is found
C *a*, 21 $\frac{3}{4}$ degrees, B C 25. degrees B *a* 11. degrees C 25. degrees
24. minutes, and *a* the Right Angle : There remains only the
angle B unknown; which you muft find by turning the Triangle,
as was taught Prob. 1. The other Right angled Spherical Triangle made, is B *a* A, wherein is found A *a* (Complement of 21 $\frac{3}{4}$
degrees to 60 degrees (the whole Bafe before given) 38 $\frac{1}{4}$ degrees, A B 38. degrees 30. minutes, B *a* 11. degrees, and *a* the
Right Angle ; which is more then enough to find the Angle A
and B; as was fhewed in the *Preface. Theorem* 1.

The End of the Sixth Book.

Here follows the *Ancient*
STORIES of the feveral STARS, *and*

CONSTELLATIONS.

Shewing the Poetical Reafons why fuch Various Figures are placed in *HEAVEN.*

Colleeted from Dr HOOD.

And Firft,

Of the Northern Conftellations.

I. URSA MINOR. This Conftellation hath the preheminence, becaufe it is neereft of all the reft unto the North Pole ; And is called of the Greeks Ἀρκτ⊙ whereupon the Pole is called the Pole Arctick, for that it is neer unto that Conftellation. It is alfo called *Helice minor,* becaufe of the fmal revolution which it maketh round about the Pole : or rather of *Elice,* a Town in *Arcadia,* wherein *Calyfto* the great Bear, and mother to the lefs, was bred. It is called *Cynofura,* becaufe this Conftellation, though it carry the name of a Bear, yet it hath the taile of a Dog : Laft of all, it is termed *Phoenice,* becaufe that *Thales,* who firft gave the name to this Conftellation, was, a *Phoenician* : And therefore the *Phoenicians* being taught how to ufe it in their Navigations, did call it by the name of the Country where-in *Thales* was born. It confifteth of 7. ftars, which the Latines call *Septemtriones;* becaufe by their continual motion, thofe feven ftars do as it were wear the Heavens. The *Spaniards* call them all *Bofina,* that is, an *Horn;* becaufe they may be very well brought into that form; whereof that which is in the end of the tail, is called the *Pole-ftar,* by reafon of the nearenefs thereof unto the Pole of the world · for it is diftant (according to the opinion of moft) from the true Pole, but 23. deg. 30. min. The Arabians
call

call it *Alrukaba* : And of the *Scythians* it is said to be an *Iron nail*, and is worshipped by them as a God. The two stars that are in the sholders of the *Bear*, are called *Guards*, of the Spanish word *Guardare*, which is to behold ; because they are diligently to be looked unto, in regard of the singular use which they have in Navigation.

The reason why this Constellation was brought into the Heavens, is diversly set down, and first in this manner: *Saturn* having received of the Oracle that one of his Sons should banish him out of his kingdom, determined with himselfe to kill all the men children that he should beget: whereupon he gave command to *Ops* his wife, being then great, that she should shew him the child so soon as ever it was born : But she bringing forth *Jupiter*, and being greatly delighted with his hair, gave the child unto two Nymphs of *Crete*, dwelling in the mount *Dicte*; whereof this was one, and was called *Cynosura*; the other was *Helice*.

Jupiter, after that (according to the Oracle) he had bereft his Father of the kingdom, in recompence of their paines and courtesie, translated them both into the Heavens, and made of them two Constellations; the *Lesser Bear*, and the *Greater Bear*.

Otherosme say that it was *Arcas*, the son of *Calysto*; and they tell the tale on this manner. *Calysto* a Nimph of singular beauty, daughter to *Lycaon* King of *Arcadia*, induced by the great desire she had of hunting, became a follower of the Goddess *Diana*. After this, *Jupiter* being enamored with her beauty and out of hope, by reason of her profession, to win her love in his own person, counterfeited the shape of *Diana*, lay with *Calysto*, and got her with child; of whom was born a son, which was called *Arcas*. *Diana*, or rather *Juno*, being very much offended here-with, turned *Calisto* into a *Bear*. *Arcas* her son at the Age of fifteen, hunting in the woods by chance lighted upon his mother in the shape of a *Bear* : who knowing her son *Arcas*, stood stil, that he might come near unto her, and not be afraid : but he fearing the shape of so cruel a Beast, bent his bow of purpose to have slain her : Whereupon *Jupiter* to prevent the mischief, translated them both into Heaven, and of them made two several Constellations : unto the *lesser Bear*, there belongs but one star unformed.

2. URSA MAIOR, the *Greater Bear*, called also of the Greeks *Arctos*, and *Helice*, consisteth of 27. stars: Among the

D d   which

which, thofe feven that are in the hinder part and tail of the Bear, are moft obferved; the Latines call them *Plauftrum;* and of our men they are called *(harles Wayn;* becaufe the ftars do ftand in fuch fort, that the three which are in the tail refemble the Horfes, and the other four which are in the flank of the *Bear,* ftand (after a manner) like the Wheels of a Waggon, or Chariot; and they are fupofed by fome to be greater then the Sun. The reafon of the Tranflation of this Conftellation into the Heaven, is at large fet down in the other Conftellation, and therefore needs not here to be repeated. This Conftellation was firft invented by *Nau-plius,* the Father of *Palamedes* the Greek: and in great ufe among the Grecians; and this is to be noted both in this and the former Conftellation, that *they never fet under the Horizon, in any part of Europe:* which though it fall out by reafon of their fcituation in the Heavens; yet the Poets fay, that it came to pafs through the difpleafure and hatred of *Juno;* who for that fhe was by *Ca-lifto* made a Cuckquean, and they notwithftanding (as fhe took it) in difpight of her, were tranflated into Heaven, requefted her brother *Neptune,* that he fhould never fuffer thofe Stars to fet within his Kingdom : To which requeft *Neptune* condifcended; fo that in all *Europe* they never come neer unto the Sea, or touch the Horizon. If any one marvel, that (feeing fhe hath the form of a *Bear*) fhe fhould have a tail fo long; Imagine that *Jupiter* fearing to come too nigh unto her teeth, laid hold on her tail, and thereby drew her up into heaven; fo that fhe of her felfe being very weighty, and the diftance from the Earth to the Heaven very great, there was great likelyhood that her tail muft ftretch.

The unformed ftars belonging to this Conftellation are eight.

3 D R A C O, the *Dragon,* of fome named the *Serpent;* of others the *Snake,* by the *Arabians, Aben;* and by *Junctinus Flo-rentinus, Vrago;* becaufe he windeth his tail round about the Ecliptick Pole; it containeth 31. ftars. This was the Dragon that kept the Golden Apples in the Orchard of the Hefperides, (now thought to be the Iflands of *(ape de Virde*) and for his di-ligence and watchfulnefs, was afterwards tranflated into heaven: Yet others fay that he came into Heaven by this occafion; when *Minerva* withftood the *Gyants* fighting againft the *Gods;* they to terrifie her, threw at her a mighty *Dragon;* but fhe catching him in her hands, threw him prefently up into Heaven, and pla-ced him there, as a memorial of that her refiftance.   Others
                                                                 would

would have it to be the Serpent *Python*, whom *Apollo* flew, after the Deluge.

4. CEPHUS, containeth in him 11. ftars, and hath two unformed. This was a King of the *Æthiopians*, and Hufband unto *Caffiopeia*, and father of *Andromeda*, whom *Perfeus* married. He was taken up into Heaven, with his wife and daughter, for the good deeds of *Perfeus* his fon in law ; that he and his whole ftock might be had in remembrance for ever. The Star which is in his right fhoulder, is called by the Arabians *Alderahiemin*; i. e. his right Arm.

5 BOOTES, *the driver of the Oxen* (for fo I fuppofe the name to fignifie, rather then an *Herdfman*; for he hath not his name becaufe he hath the care of any Cattle, but only becaufe he is fuppofed to drive *Charles his Wain*, which is drawn by 3. Oxen) he is alfo called *Arctophilax*, the *keeper* of the *Bear*, as though the care of her were committed to him. This Conftellation confifteth of 22 Stars. Some will have *Bootes* to be *Arcas*, the Son of her who before was turned into the *Great Bear*; and they tell the Tale thus : *Lycaon* the Father of *Califto*, receiving *Jupiter* into his houfe as a gueft, took *Arcas* his daughters fon, and cut him in pieces; and among other Services, fet him before *Jupiter* to be eaten: for by this means he thought to prove if his gueft were a God, as he pretended to be. *Jupiter* perceiving this heinous fact, overthrew the table, fired the houfe with lightning, and turned *Lycaon* into a wolf: but gathering, and fetting together again the limbs of the child, he commited him to a Nymph of *Ætolia* to be kept : *Arcas* afterwards coming to mans eftate, and hunting in the woods, lighted at un-awares upon his mother, transformed by *Juno* into the fhape of a *bear*, whom he perfued into the Temple of *Jupiter Lycæus*, whereunto by the law of the *Arcadians*, it was death for any man to come. For as much therefore as they muft of likelyhood be both flain, *Calyfto* by her fon, and he by the Law; *Jupiter* to avoid this mifcheif, of meer pitty took them both up into heaven. Unto this Conftellation belongeth but one ftar unformed, and it is between the legs of *Bootes*, and by the *Grecians* it is called *Arcturus*, becaufe of all the ftars neer the *great Bear* named *Arctos*, this ftar is firft feen neer her tail in the evening. The Poetical invention is thus.

Icarus the father of *Erigone*, having received of the God

Bacchus a Flagon of wine, to declare how good it was for mortal men, travelled therewith into the Territories of *Athens,* and there began to carouſe with certain ſhepheards: they being greatly delighted with the pleaſantneſs of the wine, being a new kind of liquor, began to draw ſo hard at it, that ere they left off, they were paſt one and thirty; and in the end, were fain to lay their heads to reſt. But coming unto themſelves again, and finding their brains ſcarce in good temper, they killed *Icarus,* thinking indeed that he had either poyſoned them, or at the leaſt-wiſe made their brains introxicate. *Erigone* was ready to die for grief, and ſo was *Mera,* her little dog. But *Jupiter* to allay their grief, pla-ced her father in Heaven, between the legs of *Arctophilax.*

6. C O R O N A   B O R E A, the *Northern garland,* con-ſiſteth of eight ſtars; yet *Ovid* ſaith, that it hath nine. This was the Garland that *Venus* gave unto *Ariadne,* when ſhe was mar-ried unto *Bacchus,* in the Iſle *Naxus,* after that *Theſeus* had for-ſaken her : which Garland, *Bacchus* placed in the Heaven, as a token of his love. *Novidius* will have it to be the *Crown of the Virgin Mary.*

7. E N G O N A S I S: This Conſtellation hath the name, becauſe it is expreſſed under the ſhape of a man kneeling upon the one knee, and is therefore by the Latines called *Ingeniculum.* It containeth 29. ſtars, and wanteth a proper name, becauſe of the great diverſity of opinions concerning the ſame. For ſome will have it to be *Hercules,* that mighty Conquerer, who for his 12. labours was thought worthy to be placed in the heaven, and nigh unto the *Dragon* whom he overcame. Others tell the tale thus: That when the *Tytans* fought againſt the *Gods,* they for fear of the *Gyants,* ran all unto the one ſide of the heaven: whereupon the Heaven was ready to have fallen, had not *Hercu-les* together with *Atlas* ſet his neck unto it, and ſtayed the fall: and for this deſert, he was placed in the Heaven.

8 L Y R A, the *Harp,* it containeth 10. ſtars; whereof thus goeth the Fable. The River *Nilus* ſwelling above his banks, overflowed the Country of *Ægypt;* after the fall whereof there were left in the fields divers kinds of living things, and amongſt the reſt a *Tortoiſe; Mercury,* after the fleſh thereof was conſumed, the ſinews ſtill remaining, found the ſame, and ſtriking it, he made it yeild a certain ſound; whereupon he made an *Harp* like unto it, having 3. ſtrings and gave it unto *Orpheus* the ſon of *Caſſiopea.*
                                                                                This

This *Harp* was of such excellent sound, that Trees, Stons, Fowls, and wild Beasts are said to follow the sound thereof. After such time therefore that *Orpheus* was slain by the women of *Thrace*, the Muses by the good leave of *Jupiter*, and at the request of *Apollo*, placed this *Harp* in Heaven. *Novidius* will have it to be the *Harp* of *David*, whereby he pacified the evil spirit of *Saul*. This Constellation was afterwards called *Vultur Cadiens*, the *falling Grype* : and *Falco*, the *Falcon*; or *Timpanum*, the *Timbrel*.

9 OLOR, or *Cygnus*, the *Swan*, called of the Caldaeans *Adigege*: it hath 17. stars: of this Constellation the Poets Fable in this manner. *Jupiter* being overtaken with the love of *Læda*, the wife of *Tyndarus* King of *Oebalia*, and knowing no honester way to accomplish his desire, procured *Venus* to turn her selfe into an *Eagle*, and himself he turned into the shape of a *Swan*: Flying therefore from the *Eagle*, as from his natural enemy, that earnestly pursued him, he lighted of purpose in the lap of *Leda*, and, as it were, for his more safety, crept into her bosome. The woman not knowing who it was under that shape, but holding (as she thought) the *Swan* fast in her armes, fell a sleep: In the mean while *Jupiter* enjoyed his pleasure; and having obtained that he came for, betook him again unto his wings; and in memorial of his purpose (attained under that form) he placed the *Swan* among the stars.

Ovid calleth this Constellation *Milvius*, the *Kite*, and telleth the tale thus. The Earth being greatly offended with *Jupiter*, because he had driven *Saturn* his father out of his Kingdom, brought forth a monstrous *Bull*, which in his hinder parts was like a *Serpent*; and was afterwards called the *Fatal-Bull*; because the *Destinies* had thus decreed, that whosoever could slay him, and offer up his entrails upon an Altar, should overcome the eternal Gods. *Briareus* that mighty Gyant, and ancient enemy of the Gods overcame the Bull, and was ready to have offered up his entrails according to the decree of the *Destinies* : But *Jupiter* fearing the event, commanded the Fowls of the Air to snatch them away : which although to their power, they endavoured, yet there was none of them found so forward and apt to that action as the *Kite*, and for that cause he was accordingly rewarded with a place in Heaven. Some call this Constellation *Orvis* that is, the *Bird*: others call it *Vulture volans*, the *Flying Grype*: It is also called *Gallina*, the *Hen*. Unto this Constellation do belong two unformed stars.   10. C A S

10. CASSIOPEIA, She confifteth of 13. Stars. This was the Wife of *Cepheus,* and mother of *Andromeda,* whom *Perfeus* married, and for his fake was tranflated into Heaven, as fome write. Others fay that her beauty being fingular, fhe waxed fo proud, that fhe preferred her felf before the *Nereides,* which were the *Nymphs of the Sea :* for which caufe, unto her difgrace, and the example of all others that in pride of their hearts would advance themfelves above their betters, fhe was placed in the Heaven with her head as it were downwards, fo that in the revolution of the Heavens, fhe feemeth to be carried head-long.

11. PERSEUS, he hath 26. Stars. This was the Son of *Jupiter,* whom he in the likenefs of a *Golden fhower* begat upon *Danae,* the daughter of *Acrifius.* This *Perfeus* coming unto mans eftate, and being furnifhed with the Sword, Hat, and Wings of his brother *Mercury,* and the Shield of his fifter *Minerva,* was fent by his fofter-father *Polidectes,* to kill the Monfter *Medufa,* whom he flew; and cuting off her head, carried it away with him : But as he was haftning homeward, flying in the Air, he efpied *Andromeda* the daughter of *Cepheus* and *Caffiopeia,* for the pride of her mother, bound with a Chain unto a Rock, by the Sea fide there to be devoured by a Whale: *Perfeus* taking notice and pitty of the cafe, undertook to fight with the *Monfter,* upon condition that *Andromeda* might be his Wife ; to be fhort, he delivered *Andromeda,* married her, and returning homeward unto the Ifle *Seriphus,* he found there his Grand-father *Acrifius,* whom by mifchance, and unadvifedly, he flew with a quoit : (or as *Ovid* reporteth, with the terrible fight of the horrible head of *Medufa,* not knowing that it was his Grand-father : but afterwards underftanding whom he had flain, he pined away through extream forrow : whereupon *Jupiter* his Father pittying his grief, took him up into Heaven and there placed him in that form wherein he overcame *Medufa,* with the fword in one hand, and the head of *Medufa* in the other, and the Wings of *Mercury* at his Heels. This Conftellation, becaufe of the unluckinefs thereof, is called by Aftrologers *Cacodemon, (i.e.) Unlucky,* and *Unfortunate.* For (as they fay) they have obferved it, that whatfoever is born under this Conftellation, having an evil Afpect, fhall be ftricken with fword, or loofe his Head. *Novidius* faith that it is *David* with *Goliah* his head in the one hand, and his fword in the other. The unformed Stars belonging unto this Conftellation, are three.   12. A U-

12. AURIGA, the *Waggoner*, or *Carter* : he confifteth of 14. Stars ; the *Arabians* call him *Alaiot;* the *Greeks Heniochus, i. e.* a man holding a bridle in his hand, and fo is he pictured. *E-ratoftenes* affirmeth him to be *Ericthonious* King of *Athens,* the fon of *Vulcane* : who having moft deformed feet, devifed firft the ufe of the Wagon or Chariot, and joyned horfes together to draw the fame, to the end that he fitting therein, might the better conceal his deformities. For which invention, *Jupiter* tranflated him into the Heavens.

In this Conftellation there are two other particular Conftellations to be noted ; whereof the one confifteth but of one Star alone, which is in the left fhoulder of *Auriga,* and is called *Hircus,* or *Capra* the *Goat;* the *Arabians* call it *Alhaioth* : The other confifteth of two little Stars a little beneath the other, ftanding as it were in the hand of *Auriga* ; this Conftellation is called *Hædi,* the *Kids.* The tale is thus; *Saturn* (as you heard before) had received of the Oracle, that one of his fons fhould put him out of his Kingdom, whereupon he determined to devoure them all : *Ops* by ftealth conveyed away *Jupiter,* and fent him to *Meliffus* King of *Crete,* to be nourifhed : *Meliffus* having two daughters, *Amalthæa* and *Meliffa,* committed *Jupiter* unto their Nurfery; *Amalthæa* had a *Goat* that gave fuck unto two *Kids,* fo that by the milk of this *Goat,* fhe nourifhed *Jupiter* very well. To requite this her care and courtefie, *Jupiter* (after he had put his Father out his Kingdom) tranflated her *Goat* and her two *Kids* into Heaven ; and in remembrance of the Nurfe, the *Goat* is called *Capra Amalthæa.* *Novidius* faith, that when Chrift was born, and his birth made manifeft by the Angels unto the Shepherds, one of them brought with him for a Prefent, a Goat and two young Kids; which in token of his good will, were placed in Heaven.

13. OPHIUCHUS, or SERPENTARIUS, That is, the *Serpent-bearer.* This Conftellation hath no proper name, but is thus entituled, becaufe he holdeth a Serpent in his hands. It containeth 24. Stars. Some fay that it is *Hercules,* and report the tale on this manner *Juno* being a great enemy to *Hercules,* fent two fnakes to kill him as he lay fleeping in his Cradle : but *Hercules* being a lufty Child (for *Jupiter* had fpent two daies in begetting him) without much ado ftrangled them both : In memorial of fo ftrange an event, *Jupiter* placed him in the Heavens, with a Serpent in his hands. 14. SER-

14. SERPENS, the *Serpent* of *Ophiuchus*, which confiſteth of 18. Stars. Some ſay that it is one of the Serpents that ſhould have ſlain *Hercules* in his Cradle. *Nouidius* ſaith, it is the *Viper* that bit *Paul* by the hand. Others deliver the tale in theſe words; *Glaucus* the ſon of *Minos* King of *Crete*, was by misfortune drowned in a Barrel of Honey: *Minos* his father craved the help of *Æſculapius* the Phyſitian : and that he might be driven per-force to help the child, he ſhut him up in a ſecret place, together with the dead carcaſs: whiles *Æſculapius* ſtood in a great maze with himſelf what were beſt to be done, upon a ſudden there came a Serpent creeping towards him; the which Serpent he ſlew with the ſtaff which he had in his hand. After this there came another Serpent in, bringing in his mouth a certain herb, which he laid upon the head of the dead Serpent, whereby he reſtored him unto life again. *Æſculapius* uſing the ſame herb, wrought the ſame effect upon *Glaucus*. Whereupon (after that) *Aeſcu-lapius* (whom ſome affirm to be *Ophiuchus*) was placed in the Heaven, and the *Serpent* with him.

15. SAGITTA, or *Telum*; the *Arrow* or *Dart*. This was that Arrow wherewith *Hercules* ſlew the *Eagle* or *Grype* that fed upon the Liver of *Promotheus*, being tyed with chains to the top of the mount *Caucaſus*; and in memorial of that deed, was tranſlated into Heaven. Others will have it to be one of thoſe Arrows which *Hercules* at his death gave unto *Phyloctetes*, up-on which the Deſtiny of *Troy* did depend. The whole Conſtel-lation containeth five Stars.

16. AQUILA, the *Eagle*, which is called *Vultur Volans*, the *flying Grype* : It hath in it 9. Stars. The Poetical reaſon of this Conſtellation, is this ; *Jupiter* transforming himſelf into the form of an Eagle, took *Ganimides* the *Trojan Boy*, whom he great-ly loved, up into Heaven, and therefore in ſigne thereof (becauſe by that means he performed his purpoſe) he placed the figure of the Eagle in the Heaven. There belong unto this Conſtellation 6. Stars (before time) unformed, but now brought into the Con-ſtellation of *Antinous*. But whereupon that name ſhould come, I know not, except it were that ſome man deviſed it there to cur-ry favour with the Emperour *Adrian*, who loved one *Antinous Bithynicus* ſo well, that he builded a Temple in his honour at *Mantinea.*

17. DELPHINUS, the *Dolphin*: It containeth 10. Stars;
yet

yet *Ovid* in his second Book *de Faſtis*, ſaith that it hath but nine. Neither did the ancient Aſtronomers attribute unto it any more, according to the number of the Muſes; becauſe of all other Fiſhes, the *Dolphin* is ſaid to be delighted with Muſick. The tale goeth thus concerning this Conſtellation. When *Neptune* the God of the Sea greatly deſired to match with *Amphitrite*, ſhe being very modeſt and ſhame-faced, hid her ſelf : whereupon he ſent many meſſengers to ſeek her out, among whom, the *Dolphin* by his good hap, did firſt find her; and perſwaded her alſo to match with *Neptune* : For which his good and truſty ſervice, *Neptune* placed him in the Heaven.

Others ſay, that when *Bacchus* had transformed the Mariners that would have betrayed him, into *Dolphins,* he placed one of them in Heaven, that it might be a leſſon for others to take heed how they carried any one out of his way, contrary both to his deſire, and their own promiſe. *Novidius* referreth this Conſtellation unto the Fiſh which ſaved *Jonas* from drowning.

18. E QU I C U L U S, is the little Horſe, and it conſiſteth of 4. Stars. This Coſtellation is named almoſt of no Writer, ſaving *Ptolomeus*, and *Alphonſus*, who followeth *Ptolomy*, and there-fore no certain tale or Hiſtory is delivered thereof, by what means it came into Heaven.

19. E QU U S A L A T U S, the *Winged Horſe*, or *Pegaſus*, it containeth 20. Stars. This Horſe was bred of the blood of *Me-duſa*, after that *Perſeus* had cut off her head, and was afterwards taken and tamed by *Bellerophon*, whiles he drank of the River *Pirene* by *Corinth*, and was uſed by him in the conqueſt of *Chi-mera:* After which exploit, *Bellerophon,* being weary of the earthly affaires, endevoured to fly up into Heaven : But being amazed in his flight, by looking down to the earth, he fell from his horſe ; *Pegaſus* notwithſtanding continuing his courſe, (as they feigne) entred into Heaven, and there obtained a place among the other Conſtellations.

20. A N D R O M E D A, She conſiſteth of 23. Stars; but one of them is common both unto her, and *Pegaſus*, This was the daughter of *Cepheus* and *Caſſiopeia*, and the Wife of *Perſeus:* the reaſon why *Minerva*, or *Jupiter* placed her in the Heavens, is before expreſſed. *Novidius* referreth this Conſtellation unto *Alexandria* the *Virgin*, whom S. *George* through the good help of his horſe, delivered from the *Dragon*.

E e

21. T RI-

21. TRIANGULUM, the *Triangle*, called alſo *Deltaton*, becauſe it is like the fourth letter of the Greeks Alphabet Δ, which they call *Delta*; it conſiſteth of four ſtars. They ſay it was placed in Heaven by *Mercury*, that thereby the head of the *Ram* might be the better known. Others ſay, that it was placed there in honour of the *Geometricians*, among whom, the Triangle is of no ſmall importance. Others affirme, that *Ceres* in times paſt requeſted *Jupiter* that there might be placed in Heaven ſome Figure repreſenting the form of *Sicilie*, an *Iſland* greatly beloved of *Ceres*, for the fruitfulneſſe thereof: now this Iſland being triangular, (at her requeſt) was repreſented in the Heaven under that form.

Thus much concerning the Conſtellations of the Northern Hemiſphear. Now follow the Poeical Stories of the Conſtellations of the Southern Hemiſphear.

Secondly,

Of the Southern Conſtellations.

1. CETUS, the *Whale*, it is alſo called the *Lion*, or *Bear* of the *Sea*. This is that monſtrous fiſh that ſhould have devoured *Andromeda*, but being overcome by *Perſeus*, was afterwards tranſlated into Heaven by *Jupiter*, as well for a token of *Perſeus* his manhood, as for the hugeneſs of the fiſh it ſelf. This conſtellation conſiſteth of 22. Stars.

2. ORION, this hath 38. Stars. The Poetical reaſon of his tranſlation into the Heaven, ſhall be ſhewn in the *Scorpion*, amongſt the Zodietical Conſtellations. The Ancient Romans called this Conſtellation *Jugala*; becauſe it is moſt peſtiferous unto Cattel, and as it were the very cut-throat of them. There are bright Stars in his girdle, which we commonly call our *Ladies yard*, or *wand*. *Novidius*, applying this ſword of *Orion* unto Scripture, will have it to be the ſword of *Saul*, afterwards called *Paul*, wherewith he perſecuted the Members of Chriſt: which after his converſion was placed in Heaven. In his left ſhoulder there is a very bright Star, which in Latine is called *Bellatrix*, the *Warriour*, in the foeminine gender. I cannot find the reaſon except it be this; that Women born under this Conſtellati-

on

on fhall have mighty tongues. The reafon of the *Ox-hide* which he hath in his hand, may be gathered out of the next ftory.

3. FLUVIUS, the *River*; it comprehendeth 34. Stars. It is called by fome *Eridanus*, or *Padus*; and they fay that it was placed in Heaven in remembrance of *Phaeton*, who having fet the whole World on fire by reafon of mifguiding of his father *Phoebus* his charriot, was flain by *Jupiter* with a thunder-bolt, and tumbling down from Heaven, fell into the River *Eridanus*, or *Padus*, which the Italians call *Po*. Others fay that it is *Nylus*, and that that Figure was placed in the Heaven becaufe of the excellency of that River, which by the Divines is called *Gihon* ; and is one of the Rivers of Paradice. Others call it *Flumen Orionis*, the flood of *Orion* ; and fay, that it was placed there, to betoken the Off-fpring from whence *Orion* came : for the tale is thus reported of him.

Jupiter, Neptune, and *Mercury*, travelling upon the earth in the likenefs of Men, were requefted by *Hyreus* to take a poor lodging at his Houfe for a Night : they being overtaken with the evening, yeilded unto his requeft; *Hyreus* made them good cheer, killing an *Ox* for their better entertainment : The Gods feeing the good heart of the old man, willed him to demand what he would in recompence of his fo friendly cheer. *Hyreus* and his Wife being old, requefted the Gods to gratify them with a Son. They to fulfil his defire, called for the hide of the Ox that was flain, and having received it, they put it into the Earth, and made water into it all three together, and covering it, willed *Hyreus* within ten moneths after to dig it out of the Earth again ; which he did, and found therein a Man-child ; whom he called *Ourion, ab Urina*, of pifs; although afterwards by leaving out the fecond letter, he was named *Orion*. At fuch time therefore as he was placed in Heaven, this flood was joyned hard to his heels, and the Ox hide wherein the Gods did pifs, was fet in his left hand, in memorial of his Off-fpring.

4. LEPUS, the *Hare*, which confifteth of 12. ftars. This Conftellation was placed in Heaven between the legs of *Orion*, to fignifie the great delight in hunting which he had in his life time. But others think it was a frivolous thing, to fay that fo notable a fellow as *Orion* would trouble himfelf with fo fmal and timerous a beaft as the *Hare*: and therefore they tell the tale thus.

In times paſt there was not a *Hare* left in the Iſle *Leros:* a certain youth therefore of that Iſland, being very deſirous of that
kind of beaſt, brought with him from another Country therea-
bout, an *Hare* great with young ; which when ſhe had brought
forth, they in time became ſo acceptable unto the other Countri-
men, that every one almoſt deſired to have and keep a *Hare.* By
reaſon whereof, the number of them grew to be ſo great, within
a ſhort ſpace after, that the whole Iſland became full of *Hares,*
ſo that their Maſters were not able to find them meat : whereupon the *Hares* breaking forth into the fields, devoured their Corn.
Wherefore the inhabitants being bitten with hunger, joyned together with one conſent, and (though with much ado) deſtroy-
ed the *Hares.* *Jupiter* therefore placed this Conſtellation in the
Heavens as well to expreſs the exceeding fearfulneſs of the beaſt,
as alſo to teach men this leſſon ; that there is nothing ſo much
to be deſired in this life, but that at one time or an other it bring-
eth with it more grief then pleaſure. Some ſay, that it was placed
in Heaven at the requeſt of *Ganimedes,* who was greatly delighted with hunting the *Hare.*

  5. C A N I S M A I O R, the *Great Dog,* it conſiſteth of 18.
Stars. It is called *Sirius Canis,* becauſe he cauſeth a mighty
drought by reaſon of his heat. This is the Conſtellation that gi-
veth the name unto the *Canicular* or *Dog-days* ; whoſe beginning
and end is not alike in all places, but hath a difference according
to the Country and Time : as in the Time of *Hypocrates* the
Phiſitian, who lived before the time of Chriſt 400. years, the
Canicular days began the 13. or 14. of *July.* In the time of *A-*
vicenna, the *Spaniard,* who lived in the year of our Lord 1100.
the *Canicular* days began the 15, 16, or 17. of *July.* In our
Country, they begin about *S. James-tide,* but we uſe to account
them from the 6. of *July,* to the 17. of *Auguſt* ; which is the
time when the Sun beginneth to come near unto, and to depart
from this Coſtellation.

  Novidius will have it to be referred to *Tobias* Dog; which
may very well be, becauſe he hath a tail ; *Tobias* Dog had one;
as a certain fellow once concluded, becauſe it is written that
Tobias his Dog fawned upon his Maſter, therefore it is to be
noted (ſaid he) that he had a tail. The Poets ſay, that this is the
Dog whom *Jupiter* ſet to keep *Europa,* after that he had ſtolen
her away, and conveied her into *Crete,* and for his good ſervice
                                                              was

was placed in Heaven. Others say, that it was one of *Orion* his Dogs. There belong unto this Conftellation II. Stars unformed.

6. CANIS MINOR; the *Leffer Dog;* this of the Greeks is called *Procyon,* of the Latines *Ante Canis ;* it containeth but two Stars. Some fay, that this alfo was one of *Orions* Dogs, Others rather affirm it to be *Mera,* the Dog of *Origone,* or rather of *Icarius* her father, of whom mention is made in the Conftellation of *Bootes,* and *Virgo.* This Dog of meer love to his Mafter, being flain, as is aforefaid, threw him felf into the River *Anygrus,* but was afterward tranflated into Heaven with *Origone.* Among the Poets there is great diffention which of the two fhould be the Dog of *Origone ;* fome faying one, and fome the other, and therefore they do many times take the one for the other.

7. ARGO NAVIS, the *Ship Argo,* which comprehendeth 41. Stars; this is the Ship wherein *Jafon* did fetch the Golden Fleece from *Colchis,* which was afterward placed in Heaven as a memorial, not only becaufe of the great Voyage, but alfo, becaufe (as fome will have it) it was the firft Ship wherein any man fayled. Their reafon why this Ship is not made whole is, that thereby men might be put in mind not to defpair, albeit that their Ship mifcarry in fome part now and then. Some avouch it to be the *Ark of Noe. Novidius* faith it is the Ship wherein the *Apoftles* were, when Chrift appeared unto them walking on the Sea. In one of the *Oars* of this Ship, there is a great Star, called *Canopus,* or *Canobus,* which the *Arabians* called *Shuel,* as it were a bone-fire, becaufe of the greatnefs thereof. It is not feen in *Italy,* nor in any Country on this fide of *Italy.* Some fay that *Canobus* the Mafter of *Menelaus* his Ship, was transformed into this Star.

8. HYDRA, the *Hydre;* that hath 25. Stars, and two unformed.

9. CRATER, the *Cup,* or *ftanding piece ;* that hath feven Stars. Some fay that this was the cup wherein *Tagathon,* that is, the chief God, mingled the ftuff whereof he made the fouls of Men.

10. CORVUS, the *Crow;* this hath feven Stars. Thefe 3. Conftellations are to be joyned together, becaufe they depend upon one Hiftory, which is this. Upon a time *Apollo* made a folemne feaft to *Jupiter,* and wanting water to ferve his turn, he delivered a cup to the *Crow* (the bird wherein he chiefly delighted) and

fent him to fetch water therein: The *Crow* flying towards the River, efpyed a Fig-tree, fell in hand with the Figs, and abode there till they were ripe : In the end, when he had fed his fill of them, and had fatisfied his longing, he bethought himfelf of his errand, and by reafon of his long delay, fearing a check, he caught up a fnake in his bill, brought it to *Apollo,* and told him that the fnake would not let him fill the *Cup* with water. *Apollo* feeing the impudency of the bird gave him this gift, that as long as the Figs were not ripe upon the Tree, fo long he fhould never drink : and for a memorial of the filly excufe that he made, he placed both the *Crow, Cup,* and *Snake,* in Heaven.

11. CENTAURUS, the *Centaure,* which comprehendeth 37. Stars. Some fay, that this is *Typhon,* others call him *Chiron,* the Schoolmafter of thofe three excellent men, *Hercules Achiles,* and *Æfculapius;* unto *Hercules* he read *Aftronomy,* he trained up *Achilles* in Mufick, and *Æfculapius* in Phyfick : and for his upright life he was turned into this Conftellation. Yet *Virgil* calleth *Sagitarius* by the name of *Chiron.* In the hinder feet of this Conftellation, thofe ftars are fet which are called the *Crofiers,* appearing to the Mariners as they fail towards the South Sea, in the form of a croffe, whereupon they have their name. The four ftars which are in the Garnifh of the *Centaures* Spear, are accounted by *Proclus* as a peculiar Conftellation, and are called by him *Thyrfilochus,* which was a Spear compaffed about with vine leaves: but they are called by *Copernicus* and *Clavius,* and other Aftronomers, the ftars of his Target. It fhould feem that they were deceived by the old tranflation of *Ptolome,* wherein *Scutum* is put for *Hafta, i. e.* the Target, for the Spear, as it is well noted by our Countryman M. *R. Record,* in his Book intituled *The Caftle of knowledge.*

12 LUPUS, the *Wolf,* or the beaft which the *Centaure* holdeth in his hand, containeth 19. ftars; the Poetical reafon is this, *Chiron* the *Centaure* being a juft man, was greatly given the worfhip of the Gods: for which thing, that it might be notified to all pofterity, they placed him by this beaft, which he feemeth to ftick and thruft through with his Spear, (as it were) ready to kill for facrifice.

13. ARA, the *Alter,* it is alfo called *Lar,* or *Thuribulum, i. e,* a Chimney with the fire, or a Cenfor. It confifteth of feven ftars, and is affirmed of fome Poets, to be the *Alter* where-

on

on the *Centaure* was wont to offer up his facrifice. But others tell
the tale thus. When as the great Gyants called the *Tytans*, la-
boured as much as might be to pull *Jupiter* out of Heaven, the
Gods thought it good to lay their heads together, to advife what
was beft to be done : Their condufion was, that they fhould all
with one confent joyn hands together to keep out fuch fellows :
and that this their league might be confirmed, and throughly ra-
tified ; they caufed the *Cyclops*, (which were work-men of *Vulcan*) to make them an Altar : about this Altar all the Gods af-
fembled, and there fware, that with one confent they would
withftand their enemies ; afterwards, having gotten the victory, it
pleafed them to place this Altar in Heaven, as a memorial of their
League, and a token of that good which unity doth breed.

14. CORONA AUSTRINA, the *South Garland*, it
hath 13. Stars. Some fay that it is fome trifling *Garland* which
Sagittarius was wont to wear, but he caft it away from him in
jeft, and therefore it was placed between his legs : others call
it the Wheel of *Ixion*, whereupon he was tormented for that
great courtefy he would have offered unto *Juno*, thinking indeed
to have gotten up her belly : but *Jupiter* feeing the impudency
of the man, tumbled him out of Heaven (where by the licence of
the Gods he was fomtime admitted as a gueft) into Hell, there to
be continually tormented upon a Wheel. The Figure of which
Wheel was afterwards placed in Heaven, to teach men to take
heed how they be fo faucie to make fuch courteous proffers unto
other men wives. The Greeks call this Conftellation by the name
of *Uranifcus*, becaufe of the Figure thereof : For it reprefenteth
the palate or roof of the mouth, which they call *Uranifcus*.

15. The laft is PISCIS AUSTRINUS, or *Notius*,
the *South Fifh*, which comprehendeth 11. Stars befides that
which is in the mouth thereof, belonging to the water, which
runneth from *Aquarius*, and is called by the Arabians *Fomahant*.
The reafon why this *Fifh* was placed in the Heaven, is uncertain :
yet fome affirm, that the daughter of *Venus* going into a water
to wafh her felf, was fuddenly transformed into a fifh; the which
fifh was afterwards tranflated into Heaven. The unformed Stars
belonging unto this Conftellation are fix.

*Thus much concerning the Conftellations of the Northern and
Southern Hemifphears ; now follow the Poetical Stories of the
Zodiatical Conftellations.*

                                    Thirdly

Thirdly,

Of the Zodietical Conftellations.

1. ARIES, the *Ram*, it is called by the Greeks ϵρίϚ, it containeth in it 13. Stars, which were brought unto this Conftellation by *Thyeftes*, the fon of *Pelops*, and brother of *Atreus*. This is the *Ram* upon which *Phrixus*, and *Helle* his fifter, the children of *Athamas* did fit, when they fled from their ftep-mother *Ino*, over the Sea of *Hellefpont* : which *Ram* was afterward for his good fervice, tranflated into Heaven by *Jupiter*. Others fay, that it was that *Ram* which brought *Bacchus* unto the fpring of water, when through drought he was likely to have perifhed in the defert of *Lybia*. *Novidius* will have this to be the *Ram* which *Abraham* offered up in ftead of his fon *Ifaac*. The Star that is firft in the head of the *Ram*, is that from whence our later Aftronomers do account the Longitude of all the reft, and it is diftant from the head of *Aries*, in the tenth Sphear, 27. degrees 53. minutes. The unformed Stars belonging unto this Conftellation, are five.

2. TAURUS, the *Bull*, which confifteth of 23. ftars. This was tranflated into Heaven in memorial of the rape committed by *Jupiter* on *Europa* the daughter of *Agenor*, King of Sidon; whom *Jupiter* in the likenefs of a white Bull ftole away, and tranfported into *Candia*. Others fay, That it was *Io* the daughter of *Inacus*: whom *Jupiter* loved, and turned into the form of a Cow, to the intent that *Juno* comming at unawares, fhould not perceive what a part he had playd: *Jupiter* afterward in memorial of that craftie conveyance, placed that Figure in Heaven; The reafon why the Poets name not certainely whether it be a Cow or a Bull, is becaufe it wanteth the hinder parts; yet of the moft of them it is called a *Bull.* In the Neck of the *Bull* there are certain ftars ftanding together in a clufter, which are commonly called the *feven Stars*; although there can hardly be difcerned any more then fix. Thefe are reported to be the feven daughters of *Atlas*, called *Atlantiades*, whereof fix had company with the immortal Gods, but the feventh (whofe name was *Merope*) being married unto *Syfiphus* a mortal man, did

there

herefore withdraw and hide her felf, as being afhamed that fhe was not fo fortunate in matching her felf as her fifters were. Some fay, that that ftar which is wanting is *Electra*, the eldeft daughter of *Atlas*, and that therefore it is fo dim, becaufe fhe could not abide to behold the deftruction of *Troy*; but at that time and ever fince, fhe hid her face. The reafon why they were taken up into Heaven, was, their great pittie towards their father, whofe mifhap they bewayled with continual tears. O-thers fay, that whereas they had vowed perpetual virginity, and were in danger to lofe it, by reafon of *Orion*, who greatly af-fayled them, being overtaken with their love; they requefted *Ju-piter* to ftand their friend; who tranflated them into ftars, and placed them in that part of Heaven. The Poets call them *Ple-iades*, becaufe when they rife with the Sun, the Mariners may commit themfelves to the Sea. Others will have them to be fo termed *a pluendo*; becaufe they procure rain. Others give them this name, of the Greek word □λέιονες, becaufe they be ma-ny in number. They be alfo called *Vergiliæ*, becaufe they rife with the Sun in the *Spring* time: likewife *Athoraiæ*, becaufe they ftand fo thick together. Our men call them by the name of the *feven Stars*, or Brood Hen. The Aftronomers note this as a fpecial thing concerning thefe ftars, that when the Moon and thefe ftars do meet together, the eyes are not to be medled withall, or cured if they be fore: their reafon is, becaufe they be of the na-ture of *Mars* and the *Moon*.

Moreover, there be five ftars in the face of the *Bull*, reprefen-ting the form of the *Roman* letter V, whereof one (which is the greateft) is called the *Bull's Ey*. They be called *Hyades*, and were alfo the daughters of *Atlas*, who fo long bewayled the death of *Hyas* their brother, flain by a *Lion*, that they died for forrow, and were afterwards placed in Heaven for a memorial of that great love they bare to their brother. The ancient *Ro-mans* call the *Bul's Ey, Parilicium*, or *Palelicium*; of *Pales* their goddeffe; whofe feaft they celebrated after the conjunction of this ftar and the Sun. The unformed ftars belonging unto this Conftellation, are eleven.

3. GEMINI, the *Twins*; it confifteth of 18. ftars. The Poets fay they are *Caftor* and *Pollux*, the fons of *Leda*, brethren moft loving, whom therefore *Jupiter* tranflated into Heaven. Some fay that the one of them is *Appollo*, and the other *Hercu-*

les: but the moſt affirm the former. The unformed ſtars of this Conſtellation are ſeven, whereof one is called *Tropus,* becauſe it is placed next before the foot of *Caſtor.*

4. CANCER, the *Crab,* it hath 9. Stars. This is that *Crab* which bit *Hercules* by the heele as he fought with the Serpent *Hydra* in the Fen *Lerna,* and for his forward ſervice, was placed in Heaven by *Juno,* the utter enemy of *Hercules.* In this Conſtellation, there are Stars much ſpoken of by the Poets; although they be but ſmall; whereof one is called the *Crib,* other two are the two Aſſes, whereof one was the Aſſe of *Bacchus,* the other of *Vulcan,* whereon they rode to battel, when as the Gyants made war with the Gods; with whoſe braying and ſtrange noiſe, the Gyants were ſo ſcared upon the ſudden, that they forſook the field, and fled. The Gods getting the victory, in tryumphing manner tranſlated both the Aſſes, and their manger into Heaven. The unformed ſtars of this Conſtellation are four. It is called *animal rerogradum,* for when the Sun cometh into his Signe, he maketh Retrogradation.

5. LEO, the *Lyon;* it hath 27. Stars, this is that *Lyon* which *Hercules* overcame in the wood of *Nemæa* and was placed in Heaven in remembrance of ſo notable a deed. *Novidius* ſaith, this was one of the *Lyons* which were in the den into which *Daniel* was caſt, and was therefore placed in Heaven, becauſe of all other he was moſt friendly unto *Daniel.* In the breaſt of this Conſtellation is that notable great Star, the light whereof is ſuch, as that therefore it is called by Aſtronomers Βάσιλεύς or *Regulus i. e.* the *Viceroy,* or little King among the reſt. The unformed Stars belonging to the *Lyon* are eight ; whereof three make the Conſtellation which is now called *Coma Berenices,* that is, the hair of Berenice. This Conſtellation was firſt found out and invented, by *Canon* the Mathematician, but deſcribed by *Calimachus* the Poet. The occaſion of the Story was this, *Ptolomeus Evergetes* having married his ſiſter *Berenice,* was ſhortly after enforced to depart from her, by reaſon of the wars he had begun in *Aſia* : whereupon *Berenice* made this vow, that if he returned home again in ſafety, ſhe would offer up her hair in *Venus* Temple. *Ptolome* returned ſafe; and *Berenice,* according to her vow, cut off her hair and hung it up. After certain daies, the hair was not to be found; whereupon *Ptolome* the King was greatly diſpleaſed: but *Canon,* to pleaſe the humor of the King, and to cur-

ry

ry favour with him, perſwaded him that *Venus* had conveyed the hair into Heaven. *Canon* attributeth ſeven Stars unto it, but *Ptolome* allotteth it but three, becauſe the other be inſenſible.

6. VIRGO, the *Virgin*, it hath 26. Stars. This is affirmed to be Juſtice, which among all the Gods ſomtime living upon the Earth, did laſt of all forſake the ſame, becauſe of the wickedneſs that began to multiply therein, and choſe this place for her ſeat in Heaven.

Others ſay, that it was *Aſtræa*, the daughter of *Aſtræus*, one of the Gyants that were called *Titans*, who fighting againſt the Gods, *Aſtræa* took their parts againſt her own Father, and was therefore after her death commended unto the Heavens, and made one of the 12. Signes.

Others ſay, that it was *Erigone*, the daughter of *Icarius*, who for that her father was ſlain by certain drunken men, for very grief thereof did hang her ſelf : but *Jupiter* taking pitty of the *Virgin* for her natural affection, tranſlated her into Heaven.

In her right wing there is one Star of ſpecial note, which by the Aſtronomers is called *Vindemeator* (*i. e.*) the gatherer of Graps. This was *Ampelos* the ſon of a *Satyr* and a *Nymph*, and greatly beloved of *Bacchus*, unto whom in token of his love, *Bacchus* gave a ſingular fair Vine, planted at the foot of an *Elme*, (as the manner was in old time.) But *Ampelos* in Harveſt gathering Graps, and taking little heed to his footing, fell down out of the Vine, and brake his neck. *Bacchus* in memorial of his former affection, tranſlated him into Heaven, and made him one of the principal Stars in this Conſtellation. There is another great Star in the hand of the *Virgin*, called of the Latines *Spica*, of the Greeks *Stachus*, of the Arabians *Azimech* (*i. e.*) the *Ear of Corn* : whereby they ſignify, that when the Sun cometh to this Signe, the Corn waxeth ripe. *Albumazar* the Arabian, and *Novidius*, take this Conſtellation for the Virgin *Mary*. The unformed Stars in this Conſtellation, are ſix.

7. LIBRA, the *Ballance*, it containeth 8 Stars. *Cicero* calleth it *Jugum* the *Yoak*, and here it is to be noted, that the Ancient Aſtronomers that firſt ſet down the number of the Conſtellations contained in the Zodiack, did account but eleven therein, ſo that the Signe which now is called *Libra*, was heretofore called Χελαί, that is to ſay, the *Claws of the Scorpion*, which poſſeſſeth the ſpace of two whole Signes. But the latter Aſtronomers, being de-

firous to have 12. Signes in the Zodiack, called thofe eight whereof the *Claws of the Scorpion* do confift, by the name of *Libra* ;
not that there was any Poetical Fiction to induce them thereto,
but only moved by this reafon, becaufe the Sun joyning with
this Conftellation, the Day and the Night are of an equal length,
and are (as it were) equally poyzed in a pair of Ballance. Yet (as
I remember) fome will have this to be the Ballance wherein
Juftice, called alfo *Aftræa,* weighed the deeds of mortal men, and
therein prefented them unto *Jupiter.* It hath 9. unformed Stars
appertaining unto it.

8. SCORPIO, the *Scorpion;* called of the Arabians, *Ala-
trab;* of *Cicero, Nepa.* It confifteth of 21. Stars. The Fiction is
thus. *Orion* the fon of *Hyreus* greatly beloved of *Diana,* was wont
to make his boaft, that he was able to overcome what beaft foe-
ver was bred upon the Earth: The Earth being moved with this
fpeech brought forth the *Scorpion,* whereby *Orion* was ftung to
death. *Jupiter* thereupon (at the requeft of the Earth,) tranflated
both the *Scorpion,* and *Orion* into Heaven; to make it a leffon for
ever for mortal men, not to truft too much unto their own ftrength,
and to the end he might fignify the great enmity between them,
he placed them fo in the Heaven, that whenfoever the one arifeth,
the other fetteth; and they are never both of them feen together
above the Horizon at once: *Gulielmus Poftellus* will have it to be
the *Serpent* which beguiled *Eve* in Paradife. The unformed
Stars about this *Scorpion* are three.

9. SAGITTARIUS, the *Archer.* It hath thirty one
Stars. Touching this Signe, there are among the Poets many and
fundry opinions. Some fay that it is *Crocus,* the fon of *Puphene,*
that was nurfe unto the Mufes. This *Crocus* was fo forward in
learning of the liberal fciences, and in the practife of feats of acti-
vity, that the Mufes entreated *Jupiter* that he might have a place
in Heaven. To whofe requeft *Jupiter* inclining, made him one of
the 12. Signes: And to the end that he might exprefs the excellent
qualities of the Man, he made his hinder parts like unto a Horfe,
thereby to fignify his fingular knowledge in Horfe-manfhip: and
by his Bow and Arrow, he declared the fharpnefs of his Wit.
Whereupon the Aftrologers have this conceit, that he that is born
under *Sagittarius,* fhall attain to the knowledge of many Arts,
and be of prompt wit, and great courage. *Virgil* affirmeth this to
be *Chiron* the Centaur, who for his fingular learning and Juftice,

was

was made the Master of *Achilles*. At which time *Hercules* coming to visit him (for he had heard both of the worthines of the
School-master, and of the great hopes of the Scholler) brought
with him his quiver of Arrows dipped in the blood of the Serpent
Hydra; but *Chiron* being desirous to see his shafts, and not taking
heed of them, being in his hand, let one of them fall upon his
foot, and being greatly tormented, not only by the anguish of the
poyson working in the wound, but much more because he knew
himself to be immortal, and his wound not to be recovered by
medicine, he was enforced to make request unto the Gods, that he
might be taken out of the World, who pittying his case, took
him up into Heaven, and made him one of the 12. Signes.

10. C A P R I C O R N U S, the *Goat*, it consisteth of 28. Stars.
The Poets say, that this was *Pan*, the God of the Shepherds, of
whom they faign in this manner : The Gods having war with
the Gyants, gathered themselves together into *Ægypt*, *Typhon*
the Gyant pursued them thither, whereby the Gods were brought
into a quandary, that well was he that by changing his shape
might shift for himself. *Jupiter* turned himself into a *Ram*: *Apollo*
became a *Crow* : *Bacchus*, a *Goat* : *Diana* lurked under the form
of a *Cat* : *Juno* transformed her self into a *Cow* : *Venus* into a *Fish*:
Pan leaping into the River *Nilus*, turned the upper part of his
body into a *Goat*, and the lower part into a *Fish*. *Jupiter* won-
dring at his strange device, would needs have that Image and
Picture translated into Heaven, and made one of the 12. Signes.
In that the hinder part of this Signe is like a Fish, it betokeneth
that the latter part of the moneth wherein the Sun possesseth this
Signe, inclineth unto Rain.

11. A Q U A R I U S, the *Waterman*. It hath 42. Stars,
whereof some make the Figure of the Man: other some the Wa-
terpot; and some, the stream of water that runneth out of the pot.
This is feigned to be *Ganimedes* the *Trojan*, the son of *Oros*,
and *Callirhoe* whom *Jupiter* did greatly love for his excellent
favour and beauty, and by the service of his Eagle carried him
up into Heaven, where he made him his Cup bearer, and called
him *Aquarius*. Others notwithstanding thinke it to be *Deucali-*
on the son of *Prometheus* whom the Gods translated into Heaven, in remembrance of that mighty deluge which happned in
his time, whereby mankind was almost utterly taken away from
the face of the earth. The unformed stars belonging unto this
Signe are three.   Ff3   12. P I S-

12. P I S C E S, the *Fishes:* these, together with the line that knitteth them together, contain 24. Stars. The Poets say that *Venus* and *Cupid* her son coming upon a certain time unto the River *Euphrates*, and sitting upon the bank thereof, upon a sudden espied *Typhon* the *Gyant*, that mighty and fearfull enemy of the Gods coming towards them; Upon whose sight, they being stricken with exceeding fear, lept into the River, where they were received by two Fishes, and by them saved from drowning. *Venus* for this good turn, translated them into Heaven. *Gulielmus Postellus* would have them to be the two Fishes wherewith Christ fed the 5000. men. The unformed stars of this Constellation, are four.

Thus have I breifly run over the Poetical reasons of the Constellations : It remains now that I speak of the *Milky way*.

V I A L A C T E A, or *Circulus Lacteus;* by the Latines so called; and by the Greekes *Galaxia;* and by the English the *Milkey way*. It is a broad white Circle that is seen in the Heaven, in the North Hemisphere, it beginneth at *Cancer*, on each side the head thereof, and passeth by *Auriga*, by *Perseus*, and *Cassiopeia*, the *Swan;* and the head of *Capricorn*, the tayl of *Scorpio*, add the feet of *Centaur*, *Argo* the *Ship*, and so unto the head of *Cancer*. Some in a sporting manner do call it *Watling street;* but why they call it so, I cannot tell; except it be in regard of the narrownesse that it seemeth to have; or else in respect of that great high way that lieth between *Dover* and *S. Albons*, which is called by our men, *Watling street.*

Ovid saith, that it is the great Causey, and the high way that leadeth unto the Pallace of *Jupiter;* but he alledgeth not the cause of the whitenes: belike he would have us imagine that it is made of white Marble.

Others therefore alledge these causes: *Jupiter* having begotten *Mercury* of *Maia* the daughter of *Atlas*, brought the child when he was born, to the breast of *Juno* lying a sleepe: But *Juno* awaking threw the child out of her lap, and let the milke run out of her breast in such aboundance that (spreading it self about the Heaven) it made that Circle which we see. Others say, that it was not *Mercury*, but *Hercules;* and that *Juno* did not let the milke run out of her breast; but that *Hercules* suckt them so earnestly, that his mouth run over, and so this Circle was made.

Others say; that *Saturn* being desirous to devour his children,

his

his wife *Ops* prefented him with a ftone wrapped in a clout, inftead of his child: This ftone ftuck fo faft in *Saturn* his throat as he would have fwallowed it, that without doubt he had therewithall been choaked, had he not been relived by his wife, who by prefling the milke out of her breafts faved his life: the milke that mifled his mouth (whereof you muft fuppofe fome fufficient quantity) fell on the Heavens, and running along made this Circle.

Dr HOOD *Commenting upon Conftellations, faith;* The Stars are brought into Conftellations, for Inftruction fake, things cannot be taught without names: to give a name to every Star had been troublefome to the Mafter and for the Scholler; for the Mafter to devife, and for the Scholler to remember: and therefore the Aftronomers have reduced many Stars into one Conftellation, that thereby they may tell the better where to feek them; and being fought, how to exprefs them. Now the Aftonomers did bring them into thefe Figures, and not into other, being moved thereto by thefe three reafons : firft thefe Figures exprefs fome properties of the Stars that are in them; as thofe of the *Ram* to to be hot and dry; *Andromeda* chained betokeneth imprifonment, the head of *Medufa* cut off fignifieth the lofs of that part: *Orion* with his terrible and threatning gefture, importeth tempeft and terrible effects : The *Serpent,* the *Scorpion,* and the *Dragon,* fignify poyfon: The *Bull* infinuateth a melancholy paffion: The *Bear* inferreth cruelty, &c. Secondly, the Stars, (if not precifely, yet after a fort) do reprefent fuch a Figure, and therefore that Figure was affigned them: as for example, the *Crown,* both North and South: the *Scorpion,* and the Triangle, reprefent the Figures which they have. The third caufe was the continuance of the memory of fome notable men, who either in regard of their fingular pains taken in Aftronomy, or in regard of fome other notable deed, had well deferved of Man-kind.

The firft author of every particular Conftellation is uncertain; yet are they of great antiquity; we receive them from *Ptolomie,* and he followed the *Platonicks;* fo that their antiquity is great. Moreover we may perceive them to be ancient by the Scriptures, and by the Poets. In the 38. Chapter of *Job* there is mention made of the *Pleiades, Orion,* and *Aucturus,* and *Mazzaroth,* which fome interpret the 12. Signes: *Job* lived in the time of *Abraham,* as *Syderocrates* maketh mention in his Book *de Commenfurandis locorum diftantiis.* Now

Now befides all this, touching the reafon of the invention of thefe Conftellations, the Poets in fetting forth thofe Stories, have this purpofe, to make men fall in love with Aftronomy: When *Demofthenes* could not get the people of *Athens* to hear him in a matter of great moment, and profitable for the Common-wealth, he began to tell them a tale of a fellow that fold an Afs; by the which tale he fo brought on the Athenians, that they were both willing to hear his whole Oration, and to put in practice that whereunto he exhorted them. The like intent had the Poets in thefe Stories: They faw that Aftronomy being for commodity fingular in the life of man, was almoft of all men utterly negle-cted: Hereupon they began to fet forth that Art under thefe Fi-ctions ; that thereby fuch as could not be perfwaded by commo-dity, might by the Pleafure be induced to take a view of thefe matters, and thereby at length fall in love them. For commonly note this, that he that is ready to read the Stories, cannot content himfelfe therewith, but defireth alfo to know the Conftellation, or at leaftwife fome principal Star therein.

FINIS.

A Discourse

OF THE

Antiquity, Progress,

AND

Augmentation

OF

ASTRONOMIE.

FIRST it seems not to be doubted, but that there was some kind of observation of Bodies Coelestial, as soon as there were Men : considering that the spectacle which the Heavens constantly present, is both so glorious, and so usefull, that men could not have eyes to see, and not fix them attentively and considerately thereupon. For, among other Apparences, when they saw the Sun dayly to change the places of its rising and setting ; at certain times of the Year : to approach neerer to the Earth in its Diurnal arch, and at others again to mount up to a height much more sublime and remote from it : and that his coming neerer to the Earth made *Winter,* and his remove higher made *Summer* : we say, when they beheld these things ; doubt-

Articl. I.

Observation Celestial, from the beginning of the world, though rude and in-artificial.

B

less

less, they could not but seriously remark and consider this vicissitude, according to which they might expect the Season would be more hard, or mild, to them in this lower Region of the World. Again, so admirably various did the Moon appear, in her several shapes and dresses of light, that she could not but invite mens eyes, and engage them to frequent Speculations : specially when she assumed those various faces or apparences, at set and certain Times ; in respect whereof it came to pass, that every Nation measured their times and Seasons, by those her constant and periodical circuits ; and this, because those periods succeded much more frequently, than the Erections and Depressions of the Sun. To these, we may add that beautifull shew of the Nightly Stars, undergoing likewise their Variations, according the variety of Seasons ; and more particularly that bright star of *Lucifer,* rising sometimes later, sometimes earlier, and sometimes not at all before the Sun, and the like. But, what we shall principally note, is only this ; that though Mankind was long, before they came to make inquiries into the Causes of these Coelestial changes and variations, restrained to set periods : yet they observed them from the very first Age, and not only admired, but also accommodated what they observed, to the uses of their Lives and their Successors. Here it might not be fruitless, to remember, that PROMETHEUS, who was imagined to have framed the first Man, was also imagined to have given him an erected Figure, and sublime Countenance; to the end he might the more advantageously advance his eyes to the Heavens, and contemplate the glory and motions of the Coelestial Lights. But, because this is too General, and rude a way of observation ; and it is our business to look back into those times *wherein men first made such Observations of Sydereral bodies, as gave hem the hint and occasion of reducing them to Method, and founding the principles of the Art, or science of Astronomy, thereupon :* we must have recourse to the monuments in *Sacred Writ,* for the understanding of that obscure matter.

*Articl. II.
Sacred records examined, and Moses found to be the First Astronomer there spoken of.*

And indeed, the light we expect from *Sacred Leaves,* would soon be clear enough to discuss all the darkness, wherein the Original of Astronomy seems involved ; could we but from them deduce the least evidence for that which the learned Antiquarie among the Jewes, *Josephus* affirms of the Sons of *Seth; viz. that they invented the science of the Heavens, before the Flood, and*

and engraved the ſame on two *Pillars,* the one of *Brick,* the other   *lib. I. Ant.*
of *Stone,* that ſo it might be preſerved in the one, in caſe the fury   *c. 3.*
of the *Deluge* to come, ſhould demoliſh and deface the other : or
if there remained to us any the moſt ſlender teſtimony of the
Reaſon he there gives, of the ſo great Longevity of men in thoſe   *cap. 4.*
dayes; namely, that the duration of their lives was ſufficiently
long, to perfect the knowledge of Aſtronomy, which requires full
600 Years, at the leaſt, to the obſervation of all the Varieties of
Coeleſtial motions : Whereupon He notes, that the Great Year
(as they call it) doth conſiſt of ſix hundred common Years ;
the vulgar opinion being, that the Celeſtial Motions do continually vary,

> *Donec conſumpto, Magnus qui dicitur, Anno,*   *Epigram de*
> *Rurſus in antiquum redeant vaga ſidera curſum,*   *ætat, Anim.*
> *Qualia præteriti ſteterant ab origine mundi.*

   Again, the buſineſs might be deduced from not long after
the Flood, if in Scripture we could find but the leaſt word from
whence might be argued the truth of what the ſame Author
writes; namely that the *Egyptians were taught Aſtronomy, by A-
braham.* Probable enough it is, we confeſs, that *Beroſus* and
others, quoted aſwell by *Joſephus,* as *Euſebius,* had read ſome
ſuch thing in ſome Books of the old Rabbins : but that the ſame
ſhould be fetched from Holy Writ, is moſt improbable, therein
being no mention at all of any ſuch thing. Beſides, there are
pious and learned Doctors, and among them *Salianus,* who will
not allow it to be ſo much as probable, that *Abraham* ſhould
inſtruct the Egyptians in Aſtronomy : becauſe of the very ſmal
time of his ſtay among them in Egypt. It is written indeed, that
Abraham came from Ur of the Chaldeans : but not that he re-   *Gen. II.*
ceived Aſtronomy from the Chaldeans, or that he delivered it
from them to the Egyptians. And therefore they conclude, that
what *Joſephus* ſaid of *Abrahams* reading Aſtronomy in Egypt,
may with more probability be imputed to his Great-grand child,
Joſeph. Concerning Him, therefore, we read (in truth) that
he was ſingularly favourable to the Prieſts in Egypt, at ſuch time,
as all the reſt of the people mortgaged their lands to the King,
for bread, during that wofull and long Famine. For, He excep-

       ted

Gen. 47.

ted the Lands belonging to the Prieſthood, and (as the Text ſaith) aſſigned them certain portions out of the publick Grana-ries ; ſo that from hence may be proved (what *Ariſtotel* tells us, from other Authority) that amongſt the Egyptians, the moſt ancient Nation, the Prieſts were exempted from labour, and left to the eaſy imploiment of their minds: and that this gave them occa-ſion to invent and conſtitute the Mathematiques : and yet for all this, it is not written, either that *Joſeph* taught thoſe Egyptian Prieſts the Mathematiques, or that they taught them to him. And, perhaps that Favour He ſhewed the Prieſts, was an argument not only of the Reſpect and Veneration, born them by the King and all his people; but alſo of his particular Gratitude toward them; in that He, who had been bred up only to Sheppardry and Coun-try imployments, and was wholy ignorant of all Arts and Sciences, at his firſt comming among them being afterwards ad-vanced to the height of a Courtier, and luſtre of a Favorite, had bin inſtructed by them in ſomething more noble and ſublime. And truely, the Divine *Moſes*, not long after admitted into the ſame Court, is not delivered ſo much to have erudited any others, as to have been himſelf *learned in all the Wiſdom of the Egyptians.* Nevertheleſs, conſidering that this Wiſedom of the Egyptians, doubtleſs contained the Mathematiques ; and that Aſtronomy was ever eſteemed the beſt and nobleſt part of them : this Eru-dition of *Moſes* ſeems to be the moſt Ancient monument of the Science of the Stars, that can be found in Holy Writ.

1. Polit. cap.
and 1 Metaph.
cap. 1.

Act. .7

    Aſtronomy, you ſee, is of great Antiquity, even upon the Records of Divinity ; and might be proved of much greater, could we but evince (what ſome alleage) that the Hiſtory of *Job* was penn'd by *Moſes*, as living a good while after him. Becauſe *Job* there mentions *Arcturus, Orion,* and the *Hyades,* or watery Conſtellation : and therefore it muſt be, that before that time the Stars had been ranged and diſpoſſed into certain Aſteriſms according to ſome certain method or artificial Theory then in uſe. But, be the time of his life never ſo uncertain, yet we may certainly obſerve from the Hiſtory thereof ; that it ſeems *Job*, being an Alien to the Hebrews, derived his knowledge of God from that which in Scripture is called, *Coelorum Exercitus,* the *Hoſt of Heaven.* For aſmuch, as the Inviſible things of God are not ſo well learned from any viſible things of Nature, or the effects of his Wiſedom and Power, as from the Coeleſtial Orders

Cap. 9.

Orders and therefore *Syneſius* juſtly calling Aſtronomy [ϗϖϱόϯμ- νον ὅϯιςήμίω] *a truly-venerable Science;* he ſaith, that it advanceth the mind to ſomthing of greater both Antiquity and Nobility, *viz,* ineffable Theology. That we may be breef, and only touch upon that ſentence in the Book of Wiſedom that God gave to *Solomon,* among other of Natural Science, to *underſtand the* *Courſe of the Year and the Diſpoſitions of the Stars* : if any thing in Sacred Writ doth expreſſly prove the Antiquity of Obſervations Aſtronomical, and the founding or erecting any ſetled Art there-upon; it muſt be that, of which the Holy Prophets complaind in their dayes ; *viz,* that there were Chaldæans, who at *Baby-lon, did contemplate the Stars, and compute the Months, that from them they might foretell things to come.* For, from hence we underſtand, that the obſervation of the Motions of Heavenly bodies was a certain profeſt Art; and of great Antiquity, among the Chaldeans. *de done ad Pæon.* *cap. 7.* *Eſa. 47.*

In the *Second* place, we are to revolve the Records of *Eth-nick Authors,* to ſee if among them we can find the time of the Nativity of Aſtronomy. *Articl, 3. Ethnick monu-ments likewiſe revolved ; and*

Look we therefore back, firſt, into the remains of that part of Time, which is called *Obſcure,* or *Fabulous;* becauſe poſſibly enough ſomething of truth, concerning our enquiry, may be found wrapt up in the darkſome ſhrouds of Fables. And be-gin we at the moſt ancient of Heathen Gods, *Coelus,* in Greek Ὀύϱϱνϴ who, as *Diodorus Siculus* delivers, was ſo named, becauſe of his high devotion to, and delight in the obſervation of the Stars. This eminent perſon being the Father of many Sons as *Atlas, Saturnus,* the *Titanes,* and among thoſe eſpeci-ally *Hyperion* and *Japetus* ; it is lawfull for us to conjecture, that led by his example, his whole family were addicted to the ſame Study. For ſeeing, that *Coelus* lived in *Mauritania,* not far from the Ocean ; and thence extended his Kingdom not only over all *Africa,* but alſo into a conſiderable part of *Europe* : it is well known that his Son *Atlas,* who ſucceded him in the ſame Dominions, is allowed to have given his name to the higheſt Mountain of that Country; only becauſe he had made his obſervations of the motions of the Heavens and Stars, from the top thereof. For, the Ancients in thoſe dayes, as the vul-gar now in ours, imagined the arch of the Heavens to be ſo little diſtant from the tops of great Mountains ; as that by how *firſt thoſe of Fabulous times : accor-ding to which Coelus is found the moſt anci-ent Aſtrono-mer : lib. 3.* *and after him his Sonns.* *1.* *Atlas, who taught Aſtro-my to his Son.*

　　much

much the higher any man afcended on thofe hills, by fo much the more clearly and diftinctly might he behold Coeleftial objects. To this, *Diodorus, Plinie*, and others add ; that *Atlas* was feigned to fupport Heaven on his Shoulders, only becaufe He had framed a Sphear, wherein the whole Heavenly machine was ftrongly reprefented : and *Clemens Alexandrinus* obferves, that *Hercules*, being both *Vates* and *Phyficus*, a Prophet and Philofopher, was reported to come and relieve *Atlas* (his great Uncle) by taking the vaft Burden of Heaven upon his own Shoulders ; becaufe He fucceded him in that difficult task the Study, or fcience of Coeleftial bodies. Of *Hefperus*, the Sonne of *Atlas*, it is recorded, that while he was bufy in fpeculating the Stars, on the top of the fame mountain, he was fnatched away by the violence of fome difeafe, and could never be found: and that thereupon, the common people, in refpect of his piety and juftice, gave his name to the moft beautifull and refplendent Star, which is alfo called *Vefperugo*, being *Venus*, while fhe is in the *Weft*. As for his Sifters, called both *Atlantiades*, and *Pleiades* ; thefe likewife gave their name to that glomeration of Stars, which are vifible in the back of *Taurus* : and of one of them, named *Maia*, was born the *Famous Mercurius*, faid to have brought the Science of the Stars firft into *Egypt*. Whence *Marcilius*, writing of the Aftronomy of the *Egyptians*, Saies of *Mercury*

Tu Princeps, Authorq, facri, Cyllenie, tanti. &c.

Though we well know, that the *Ethiopians*, allowing the *Egyptians* to be no other, but one of their *Colonies*, fent abroad to find room to fubfift in, contend; that they recieved *Aftronomy* from them: as firft *Diodorus*, and after *Lucian* have obferved. Here it is well worthy our commemoration, what *Cicero* faith, as of *Atlas* and *Promotheus*, fo alfo *Cepheus*, a King of
” the *Ethiopians* : viz. Neither had *Atlas* been beleived to
” have fuftain'd Heaven, nor *Prometheus* to have been chain'd on
” *Caucafus* : nor *Cepheus* with his Wife, Son-in Law, and Daugh-
” ter, to have been Stellified : had not their Divine cognition
” of Coeleftial bodies firft occafioned the perpetuation of their
” Names in the difguife of Fables.

To

lib. 3.
lib. 2. cap. 8.

Aftrom. 1.

Hifperus.

And Daughters,
the Atlantides
and Pleiades,
from one of
whom came
Mercury.
1. Aftron-

lib. 3.
lib. de Aftrol.
5. Tufculan.

To return to *Saturn,* another *Sonne* of *Coelus* ; He, leaving *Africa,* and reigning only in *Italy, Sicily* and *Crete;* may be thought to have profecuted his Fathers ſtudies, no leſs than the former : and we have this argument for it, that the Sloweſt of all the Planets bears his name, to this very Day ; probably, bacauſe he was the firſt, who underſtood the motion and courſe of that Planet, which was by the *Greeks* called Κρόϑ⊙ from χρόν⊙ Time, foraſmuch as of all the Coeleſtial Circuits none was found ſo diuturn. And of his Sons, ſince *Pluto* addiᶜted himſelf intierly to Huſbandry, *Neptune* to Navigation ; we may conceive, that *Jupiter,* applying his mind to nobler cares, ſucceded his Father in the Study of the Heavens : as alſo that he choſe *Olympus,* accounted the higheſt mountain, to make his obſervations upon : ſo that in proceſs of time, he came therefore to be called *Olympius;* and the name of that mountain to be transferred upon Heaven it ſelf; whoſe orders and laws He well underſtanding, was thereupon ſaid to have the Dominion of Heaven. Certain it is, that the *Grecians* aſcribed the Original of this nobleſt Science, partly to the *Gods* themſelves, and partly to ancient *Hero's* : which *Achilles Tatius* ſeaſonably alluding unto, introduceth old *Æſchylus* attributing to God, that He ſhewed the riſings and ſettings of the Stars, and diſtinguiſh't Winter, Summer, and the other Seaſons ; and *Ovid* Fathers the ſame wholly upon *Jupiter.*

2.

Saturn, who delivered the same to his Son.

Jupiter.

Iſagog. ad Phæn.

> *Perq̄ Hiemes, Æſtusq̄, & in æquales Autumnos,*
> *Et breve Ver, ſpatijs exegit quatuor Annum.*

1. *Metamorph.*

Beſides, it is in the Fiᶜtion, that *Jupiter* took his Father, *Saturn,* bound him, and precipitated him into Hell. Now this ſeems to intimate, that *Jupiter* having impoſed his own name upon one of the moſt eminent and illuſtrious of the Planets, gave that of his Father to another of them, that was more remote, ſituate in the deepeſt part of the Ætherial ſpaces, and of the ſloweſt progreſs : though all this while we are not ignorant, that thoſe names were fixed upon thoſe *Planets* a long time after : ſince more anciently the *Planet Jupiter* was called *Phaeton,* and that of *Saturn, Phoenon.* For, we may colleᶜt very neer as much from *Lucian,* who by *Tartarus* underſtands the immenſe Altitude

tude, or Profunditie of the Ætherial Region: & so denies that *Saturn* was either exil'd by *Jupiter* into Hell, or cast into bonds; as common heads were perswaded to beleeve.

3.
Hyperion.

As for *Hyperion* ; *Diodorus* hath a tradition, that he being of the progeny of old *Coelus*, demonstrated the courses of the Sun and Moon : and therefore called the Sun, *Helios*, after the name of his Sonne; and the Moon, *Selene*, after that of his Daughter.

4.
Japetus:
from whom
came Prome-
theus, *who*
followed the
same study.
in Eccles. 7.

Last of all comes *Japetus*, who also was the Sonne of *Coelus*, but performed nothing worthy commendation in the advance of his Fathers Speculations : but *Promotheus*, whom he begat, was therefore imagined to have been chained on the hill *Caueasus*, and to have his heart perpetually torn by a hungry Eagle or Vultur ; Because (as *Servius* expounds the riddle) with restless care, and solicitude of mind, he constantly excruciated himself with observing the Stars, and studying their Ascensions and Declinations. We shall not insist upon what follows in the same Author, namely that this *Prometheus* was the first, who introduced *Astrology* to the *Assyrians* (not far from *Caucasus:*) it being more usefull for us now to observe, that He was imagined to have stolen Fire from Heaven, for the inanimation of Man, for no other reason, but because he infused this Heaven-fetch't Knowledge into the breasts of men, and inflamed their souls with the desire and love thereof. For, as to the remainder ; for as much as *Belus* was the same with *Jupiter*, among the *Assyrians*, as *Diodorus* testifies : it is He rather, who was accounted both the most sacred of their Dieties, and the Inventor of this Sideral Science ; as *Pliny* affirms.

lib. 2.
lib. 37, c. 10.

so did Phae-
ton.

It is not needfull for us here to examine many other of the ancient Traditions, accounted likewise among the Fabulous ; as, in particular, the Fable of *Phaeton*, which hath this Mythology, that in his life time he had made a considerable progress toward the discovery of the Suns Annual course; but dying immaturely, he left the Theory thereof imperfect. That other of *Bellerophon*, whom Interpreters maintain to have been carried up to Heaven, not by a flying horse, but a studious and contemplative mind, eager in the the quest of Syderal mysteries. That of *Dædalas*, who indeed, by the same towring speculations, as by the artifice of wings mounted up to the Northern part of Heaven ; while his less ingenious Sonne, *Icarus*, falling short in his attempt of imitating his Fathers

Dædalus.
Icarus.

thers

thers ſublime flight (as not ſo well underſtanding the demon-
ſtrations of the reaſons of his Theory) flaggd very low in his
Studies : and fell from the true and apodicticall cognition of Coeleſtial motions and viciſſitudes : with many other the like, recounted by *Lucian* ; as that of *Endymion,* the favourite of the
Moon; of *Tireſias,* the Prophet, &c. Yet one thing there is, *de Aſtrol.*
mentioned as well by *Lucian,* as *Tatius;* which we cannot well *in Iſagog.*
paſs by ; which falling under the account of Heroicall times,
ſeems to come ſomwhat neer to that which is called Hiſtoricall.
And that is the notable Centention that aroſe betwixt *Atreus* Atreus *and*
and *Thyeſtes* about ſupreme dominion. For when by the pub- Thyeſtes.
like Conſent and Vote of the *Argives,* the Kingdom was to be his
of the two, who ſhould give the moſt eminent teſtimony of Science : it came to *Atreus* ſhare to be King : becauſe, though
Thyeſtes ſhowed them the ſigne *Aries,* in Heaven (for which
he was honourd with a golden Ram) yet had *Atreus* declared a
thing more excellent : while diſcourſing about the variety of the
Suns riſing, he made it appear, that the Sun and the World (*i. e.*
the Starry Orb) were not carryed the ſame, but quite contrary
wayes ; and conſequently, that that part of the Heavens which
was the *Weſt* or *Occident* of the Starry Orb, was the very riſing,
or *Orient* of the Solary. Hence that verſe of *Euripides,*

Δείξας γὰρ ἄϛρων τὼ ἐναντίαν ὁδόν
Qui Aſtrorum enim contrariam oſtendi viam.

　To the ſame times likewiſe are we to refer the *Inſtitution of* Hercules *and*
the *Olympick Games, by Hercules* ; which after a long inter- Iphitus.
ruption were renewed by *Iphitus.* For, inaſmuch as thoſe ſports *cap. 18.*
were inſtituted for no other end (as may be aſſured from *Cen-*
ſorius) but that their celebration might put men in mind of that
Intercalation of a month and half, that was to be made conſtantly every fourth Year, in reſpect of thoſe four times eleven,
or 44. Dayes, by which the moion of the Moon anticipated
that of the Sun ; and the four times ſix hours, or one whole
Day, by which the circuit of the Sun exceded 365 Dayes : manifeſt it is, that *Hercules* could not underſtand this, without having firſt exactly obſerved the Motions of Sun and Moon. Hi-
ther alſo belongs that which is reported of *Orpheus,* who muſt Orpheus.
needs have attentively obſerved the ſeven *Planets,* if it be true, as
　　　　　　C　　　　　　　　　　　*Lucian*

de Astrol.   *Lucian* averrs, that he reprefented their Harmony by his *Seven-stringed Harp* : which the *Grecians* thereupon defigned in Heaven, by fome Stars, that to this Day retain the name of

Palamedes.   *Lyra.* So likewife doth what *Sophocles* faith of *Palamedes,* who pointed out the feveral Afterifms, and particularly

Ἄρκτε Στροφάς τ᾽, ἠ κύνος ψυχρὸν δύσιν

Vrfum volutam, gelidum & occafum Canis.

Homer.   And laftly, what *Homer* recounts, that in thofe times were well known (befides *Bootes* and the *Bear*, or *Wain*)

Odyſſ. E.   Πληιάδας, θ᾽ Ὑάδας τ᾽, τὸ ϑένϑ᾽ Ὠρίωνϑ᾽.

Pleiades, atqp Hyades, roburqp ipfum Orionis.

Articl. 4.   We have now ftruggled through the Darknefs of Fabulous Times, and are advanced as far as to difcerne the twilight of

Secondly, thofe Historical times ; accor- ding to which the antiquity of Astron. Obfer- vations belongs either to the Egyptians, or Babylonians. in Epinom.

Hiftoricall. An here, the firft thing we clearly perceive, is that the *whole Controverſy about the Antiquity of Aftronomical Ob- ſervations, lies betwixt the Egyptians and the Affirians, or Baby- lonians.* For, as to the *Grecians,* though fome have thought they might put in alfo for a claim to the honour of being the An- thors of this admirable Science; yet by the Verdict even of *Plato* himfelf, they are to lay by the pretence of Competition, ,, For, fayth He, the firft who made Syderal infpections, was a ,, *Barbarian;* a more ancient Nation than ours bred thofe men, ,, who firft devoted their minds to that Study, in refpect of the ,, Summer-like ferenity and perfpicuity of the Air, fuch as *Egypt* ,, enjoyes, and *Syria,* where all the Stars are, clearly vifible and ,, no Clouds or Mifts to obfcure the beautifull face of Heaven. . And certainly, if we except what we newly mentioned, the Infti- tution of the *Olympick Games* by *Hercules,* and the reftauration of them after fome Intermiffion, by *Iphytus,* which hapned about 800. Years before Chrift; and fome places in the writings of *Homer,* and more efpecially of *Hefiod,* who lived neer upon the fame time, or not long before ; we fhall find that the *Grecians* can produce no Monuments of their Obfervations of the Hea- vens, more ancient than thofe of *Thales,* who flourifht, full 600. Years before Chrift ; and who yet borrowed his knowledge of
ÆEtherial

Ætherial Matters from *Egypt*. It being manifeſt therefore, that the *Ægyptians*, or their Prieſts, are the only men, that ought to be admitted to a Diſpute with the *Aſſirians* or *Babylonians*, or their Philoſophers, concerning the Antiquity of Obſervations ; and that their ſeveral Pleas ſeem equally reaſonable. Truly, it is no eaſy matter to determine the difference, ſo as to place the Lawrell on their heads to whom it doth of right belong. For, albeit *Joſephus* aſſignes the Honour to the *Chaldæans*, and others again ſtand firmly for the *Egyptians*: yet *Plato*, *Diodorus*, *Lucian*, *Achilles Tatius*, and others alleadge ſuch quotations for each party, as ſeem to have no other, but the authority of the parties themſelves. Nor ought that to ſeem ſtrange ; ſince both ſides equally alleadge the convenience of their vaſt Companies, and the ſerenity of the ſky ; ſince they both boaſt themſelves the Original Nation and allow their Competitors to be only *Colonies* ; ſince both glory in fabulous beginnings, which we cannot trace or diſcuſſe; and both recur to egregious falſhoods about the time when their Anceſtors firſt made Coeleſtiall Scrutenies. For, the *Chaldæans* (as we find on the Regiſter of *Diodorus*) affirm, that their Nation applied themſelves to theſe Studies, from times of incredible Antiquity *viz.* of four hundred and three thouſand Years : And the *Egyptians* (as *Cicero* obſerves) talk of Obſervations of four hundred thouſand and ſeventy Years ſtanding. Unleſs you ſhall pleaſe to conſigne the Victory to the *Egyptians*, becauſe they put a value only themſelves by Auction. As if it were not enough for them to boaſt thoſe four hundred nine thouſand Years (mentioned by *Lærtius*) in which from the time of *Vulcan*, the Sonne of *Nilus*, to that of *Alexander* of *Macedon*, there hapned of *Eclipſes* of the Sun three hundred ſeventy three, and of the Moon eight hundred thirty two. Theſe conſiderations premiſed, we cannot indeed deny, but the *Egyptians* had ſome Obſervations, ſome ages before *Thales* and other *Grecians* travelled among them : but, when we would enquire more preciſely into the time, when thoſe Obſervations firſt begun ; we find our ſelves at a loſs, and brought back again into the cimmerian obſurity of Fabulous Times.

Now foraſmuſch as, though *Pliny* writes, that *Epigenes* found no Obſervations among the *Babylonians* of above ſeven hundred and twenty Years antiquity, and thoſe engraven on artificial Tyles or Slates; and the moſt ancient *Eclipſes* deduced from them, were

1. Antiq. 8. in Epinom. 2. lib. de Aſtrolog. Iſagog.

loc. citat.

de Divinat.

in præfat.

lib. 9. c. 55.

in Almagest.
lib. 4. cap. 6.

tranſmitted to *Ptolomy,* about the ſame Number of Years before Chriſt : and that by the great *Hipparchus.* And to the ſame Time belongs what *Beroſus* and *Critodemus* ſay, that in their Dayes, there were extant no Obſervations of more than four hundred and thirty Years as may be found alſo in *Pliny :* foraſmuch, we ſay, as we have brought ſome conſiderable Monuments of Obſervations much elder than that time; yet ſhall not concede beyond what the *Chaldeans* themſelves profeſt, when they teſtified to *Calliſthenes* (who went to them upon no other errand, by

in lib. 2. de Coe
lo, and com-
ment. 46.

the perſwaſion of *Ariſtotel,* as *Simplicius* relates) that they had nothing of that kind among them beyond a thouſand nine hundred and Three Years paſt : which Years ſeem to commenſe at *Ninus,* the Sonne of *Belus,* and firſt King of the *Aſſyrians.* It is clear, that the Antiquity of Obſervations ariſeth to (but not above) one thouſand and ninety Years before *Alexander* the Great.

Artid. 5.
Yet neither of
of them obſer-
ved any thing
conſiderable ;
as to the deſig-
nation of
Times : but
corrupted what
they had obſer-
ved, to the in-
troduction of
Aſtrology Ju-
dicial.

But, alas! after all this great adoe, *What did the Obſervations themſelves amount to?* Why truely, for ought we can gather from all that is extant concerning them, thoſe of the *Ægyptians* amounted to nothing at all : and thoſe of the *Chaldeans* to very little. For the *Egyptians,* we confeſs, are ſaid to have obſerved the riſing of the Dog-Star, and ſome other, no very difficult apparences : but we have no remains delivered down to us, of that or any other particular they obſerved, with the exact deſignation of the Time, as they ought. And from the *Chaldæans* we have as little, beſides thoſe *Eclipſes* mentioned in *Ptolomy.* But, when I ſpeak of the *Egyptians,* I except *Ptolomy* himſelf and ſome others, who lived and ſtudied at *Alexandria,* about there hundred Years before the Nativity of our Saviour ; or after *Alexander :* as *Timocharis, Eratoſthenes, Hipparchus :* for all theſe were either *Grecians,* or to be accounted among *Grecians,* in reſpect of the language they uſed and wrote in, rather than among the ancient *Egyptians,* by whoſe Inventions even *Ptolomy* himſelf (one of their own Country men, without diſpute) was very little, or nothing at all aſſiſted in his Study of Aſtronomy. But, what concerns aſwell the *Egyptians,* as *Chaldeans ;* their Obſervations are to be diſtinguiſht (according to the diviſion vulgarly received) into (1.) *Aſtronomical,* and (2) *Aſtrological :* the former relating to the Motions, Magnitudes, Diſtances, and proportion of the Stars ; the Latter to the Effects of them, which they conjectured were dependent on the Vertues

and

and Influences of Heavenly Bodies, aſwell in the affections of the Air, as in the actions and affairs of Mankind. For, both Nations being wonderfully prone to Superſtitition, and ſurpriſed with exceſs of Admiration at the *Eclipſes* of the Sun and Moon, when they firſt beheld them; and obſerving ever now and then ſome Stars that moved in Courſes contrary to the *Weſt*, they began preſently to think, that thoſe apparences hapned not without natural Cauſes; and that it remained only on Mans part, to Study how thoſe events might come to be fore-known, which thoſe apparences did portend. Hereupon, having attributed the moſt powerfull Vertue to the five wandering Stars (as *Diodo-* *loc. citat.* *rus* teſtifieth particularly of the *(haldæans)* as underſtanding them to be the Proclaimers of the will and purpoſes of the Gods; becauſe they ſometimes aroſe, and ſometimes ſet in various places of the Heavens ; becauſe they varied their magnitude and colour: therefore they conceived, that they ought to adreſs their Studies and Diſquiſitions principally to theſe varieties. And, becauſe they imagined, that the higher the place was, from whence they ſhould obſerve theſe Wandering Stars, the more clearly and diſtinctly might they be diſcerned; they builded Structures of vaſt altitude ; and particularly that immenſe Tower at *Babylon*, deſcribed by *Herodotus*, from the higheſt area whereof (where *lib. 1.* ſtood alſo the Temple of *Belus*) they might exactly behold and obſerve the riſing and ſetting of the Stars, and other Syderal occurrences They took notice likewiſe, that thoſe five *Planets* did keep almoſt the ſame Courſe, as the Sun and Moon ; and thereupon they pointed out the Zodiack, imagining that there muſt be ſome eminent vertue in that part of the Heavens, becauſe all the *Planets* kept conſtantly to it. And this Zodiack they divided into 12. parts, or Signes ; becauſe the Moon run it over 12. times, and the Sun only once, in one Year : and according to the number of the Dayes, during which the Sun was in paſſing through one Signe, they diſtinguiſhed each Signe again into 30. parts, which we call Degrees. I ſhall not recount to you, how they would have Twelve Principal Dieties belonging to theſe 12. Signes, whereof each had his particular regiment over his proper Signe and Moneth dependent thereupon : nor how they ſubſtituted thirty of the fixt Stars, to aſſiſt the *Planets*, and called them *(oncelling Gods* : nor how they placed 12. Stars always viſible in the North, for goverment of

the

the Living; and as many more in the South, alwaies vifible, for the government of the Dead, there gathered together ; with many other the like dreams and ridiculous abfurdities. But the thing I think moſt worthy your notice, is, *by what rude kind of* *ad Aſtrolog.* 1. *artifice they diftinguiſhed the Zodiack into 12. Signes* ; as we *in Somn.* 21. find it defcribed, concerning the *Chaldæans,* by *Sextus Empiricus,* and concerning the *Egyptians,* by *Macrobius.*

The manner this. They took a veſſell with a ſmall hole in the bottom, and filling it with Water, ſuffered the ſame to diſtill Drop by Drop into another Veſſell, placed beneath to receive it ; and this from the moment of the riſing of ſome one Star or other, obſerved in one Night, untill the Moment of its riſing again the next Night following. The Water fallen down into the Receiver, they divided into 12. equal parts, and having two other ſmaller veſſells in readineſs, each of them fit to contain one twelfth part of the Water, they again poured all the Water into the upper Veſſel, and ſtrictly marking the riſing of ſome one Star in the Zodiack, they at the ſame Moment gave the Water leave to diſtill into one of the ſmaller Veſſells, and ſo ſoon as that was filled, obſerving likewiſe another riſing Star, they put under another ſmall Veſſell; and ſo alternately ſhifting the ſmall veſſels, they noted, if not in one Night, yet in many, the twelve Stars, by which they might diſcriminate the whole Zodiack into twelve equal parts. Now with what Art and exactneſs theſe Ancients meaſured out the Heavens, may be conjectured from this one example. I might adferr another foppery of the *Chaldæans,* from the ſame *Empiricus,* who relates ; that taking it for grant-ed, that the future fortunes of Men did depend on their particu-lar Horoſcope, or Signe riſing at their Birth; when they had a mind to divine in this Kind; Two of their wiſe men agreed toge-ther in the calculation of the Nativity of the Perſon propoſed: the one ſtood by the Mother in Travell, the other on ſome high place neer at hand ; and as he that was below gave the Signe, that the Infant was then newly come into the World, the other above *ibid. cap.* 20. took care to obſerve the Signe, that was juſt then newly riſen. But, it will be of more uſe for us to hear what *Macrobius* tells of the *Egyptians.* They, when they would know the Diameter of the Sun, had in readineſs a Veſſell of Stone, hollowed to the form of an Hemiſphere, exactly made, with a ſtyle or Gnomen erected in the middle, and twelve Horary Lines drawn within.

And

And on the very Day of the Equinox, obſerving the Moment, when the upper Limbus of the Sun firſt ſhewed it ſelf above the Horizon; they marked that place on the brim of the Veſſel, on which the Gnomen caſt its ſhadow. Then again marking that place on which the ſhadow ended, when the lower Limbus of the Suns body appeared juſt above the edge of the Horizon ; they meaſured the ſpace or diſtance betwixt the two marks of the Shadowes, and found it to be the ninth part of an hour, or the hundred and eight part of the Hemiſphere, and conſequently the two hundred and ſixteenth part of the whole Circuit : and from thence they deduced, that the Diameter of the Sun was the two hundredth and ſixteenth part of its whole Orb; (which, in truth, is the 700th neer upon) or did contain one Degree and an hundred Minutes ; which yet is no more than halfe a degree, or 31. Minutes, at moſt. To this we might ſuper-ad, that it was the practice of elder times, to commenſurate the Diameter of the Sun by an Hydrologie, or Veſſel of Water; collecting the ſame from part of the Water flowing down the whole Day, which had dropped untill the Sun was wholly riſen; as is inſinuated by *Plutarch*, and deduced from *Capella* : but *Cleomedes* hath at large declared, that this way of meaſuring by Water falling ſlowly and equally from Veſſel to Veſſel, was an Invention of the *Egyptians*. Now the reaſon, why we touch upon theſe particulars, was only to ſatisfy, that (as we ſaid afore) no great matter in Aſtronomy was ever obſerved either by the *Egyptians*, or by the *Babylonians*.

And, if you deſire any further Argument thereof ; Pray take this. They were very far from ſuſpecting that the Fixt Stars had any motion proper to themſelves ; or that they had any Eccentricity (excepting only that the *Egyptians* thought *Venus* and *Mercury* to move round about the Sun, as their center ; as is affirmed by *Macrobius*, and ſome others) or that the Sun had any *Apogæum* at all, with many other Particulars fully as conſiderable. Which doubtleſs muſt be the reaſon, why they invented no *Hypotheſes*, by which they might regulate themſelves, in making their Caluclations of the various motions of the heavenly bodies. And *Peter Ramus* not long ſince complaind, that we have not our *Aſtronomy* free from the trouble of *Hypotheſes*; ſuch as the Interpreters of *Ariſtotel* themſelves, and *Proclus* on *Timæus* have recorded the *Egyptians* and *Babylonians* to have
had

had amongſt them : while, incroth, he complaind, that we had not our *Aſtronomy* as rude wild and imperfect, as theirs was. For, however ſome *Hypotheſes* are more ſimple (and ſo more eaſy) than other-ſome : yet it is abſolutely impoſſible, that *Aſtronomy* ſhould conſiſt without ſome or other. Hereupon, they could obſerve, indeed, that the Planets were one while Direct in their Progreſs, another while Retrogarde, and then again Stationary ; that they in their wanderings ſometimes inclined towards the *North*, and ſometimes deflected again toward the *South* : but all that while, they could neither comprehended the reaſons of thoſe various apparences, nor calculate them by numbers. The moſt they could doe, was darkly to repreſent thoſe motions, by certain *Hieroglyphicks*, as in particular by the windings and flexures of Serpents; and the motion of the Sun, by only a Beetle rowling his pill of dung backward : as we may read in *Clemens Alexandrinus* : and then came *Eudoxus*, who having learned that variety of motions among them, was the firſt who invented *Hypotheſes* of various Orbs, for the Solution of the *Phenomena*.

Again, they were very far from attaining the determinate places of the Fixt Stars, according to Longitude and Latitude ; or according to their Right Aſcenſion, and Declination : ſo that neither could they define the true places of the Planets, by Comparation to the Fixt Stars, nor (conſequently) deſigne any Obſervations with due exactneſs. And truely, this was the Cauſe why *Hipparchus* met with no Obſervations, either of the *Egyptians* or *Babylonians*, by which he could receive the leaſt help or advantage, toward his compoſing either *Hypotheſes*, or Tables, to repreſent the motions of the Five errant Stars : and *Ptolomy* was the firſt, who partly by the benefit of Obſervations left him by *Hipparchus*, and partly by thoſe he made himſelf, became able to attempt ſuch a Work ; as ſtands recorded in his *Almageſt*. There were only the *Eclypſes*, which both theſe Nations had ſet down : as obſerved in their Commentaries: and thoſe only ſo, as that from Paſt, they might be able to conjecture ſomthing of what were to Come. Not from the motions of Sun and Mon, exactly calculated by the help of Tables ; but having learnd from common experience, that every ninetneenth Year, *Eclypſes* did return again upon the ſame Day, for the moſt part : thereupon they endeavoured to prædict what *Eclipſes* would happen, and the time when; and this after they had perceived not any

Ano-

Anomaly in the Sun, but ſome certain Inequality in the Moon, which reducing to a medium, they concluded that the Moon did every Day run through thirteen Degrees, and a little more than one ſixth part of a degree ; as *Geminus* delivers of the *Chaldæ-ans*. But in their predictions of *Lunar Eclipſes,* they were ſomwhat more confident ; aſwell becauſe theſe *Eclipſes* uſually uturn, for the three Ages next ſucceding, within the compaſs of the ſame Dayes ; as becauſe it is very rare, in reſpect of the greatneſs of the Earths ſhadow, but the Moon, either in the whole, or ſome part of her, more or leſs, falls into it : but, becauſe (as to *Solary Eclypſes*) the Moon is both ſo ſmall, and hath ſo large a Parrallax, as that ſhe doth not for the moſt part intercept the light of the Sun from the Earth ; therefore was it (as *Diodorus* witneſſeth ſpecially of the *Babylonians*) that they durſt not determine *Eclypſes* of the Sun to come, to any certaine time; but if they predicted any, with limitation of time, they always (to ſave their credit, in caſe of failing) annexed this Condition, *If the Gods be not prevailed upon, by Sacrifices and Praiers, to avert them.*

Truth is, theſe *Aſtronomers* were alſo Prieſts, and it was their intereſt to caſt in this Proviſo. For, being ambitious to be reputed interpreters of the Will of the Gods to the People, and ſo both knowing in things to come, and ſkillfull in ſuch Ceremonies, wherewith their reſpective Deities were moſt attoned and delighted : unwilling to be thought able to predict nothing, and as unwilling again to be found erring in their chief predictions, they wrapt up all in Miſteries, and amuſed the vulgar with ſuperſtitious opinions and rites. The *Egyptians,* in a great part of their ſacred Worſhip, had recourſe to the *Aſtrological* Books of their *Mercurius* (one of the Order of the Fixt Stars; a ſecond, of the Conjunction of Sun and Moon ; a third and fourth, of their riſing.) which with what ceremonious Pomp they uſed to carry about with them, in a kind of ſolemne Proceſſion, you may find amply deſcribed by *Clem. Alexandrinus.* Nor is it *lib. 6. ſtromat.* ſtrange that thoſe Prieſts accounted ſo ſacred and knowing, ſhould alſo be eſtemed for Prophets. Further, you meet with no mention of the Five Errant Stars, all this while ; and the reaſon ſeems to be, becauſe they attributed an energie of them only as they were referrable to the Inerrant or Fixt, and particularly, as they poſſeſt this or that part of ſome Signe in the Zo-

D diack

diack, and together with it had their rising, or setting. For, so much did they ascribe to the Zodiack, as that the *Babylonians*, and (in imitation of them, the *Persians* and *Indians*) thought, that each decimal of degrees, or thirds of the Signes, (and the *Egyptians* came as low as to each single degree) could not be varied in the rising, but some eminent variation most happen, especially in him, who should be borne at that time. And hereupon was it, that the *Egptians* made that great Circle of Gold (described in *Diodorus*) of a cubit in thickness, and three hundred sixty five cubits in circumference (plundred at last by *Cambyses*) that upon each cubits space might be inscribed each Day of the Year, 365. Dayes in the whole round, and also what Stars did rise, what set upon each Day, nay the very hour of their respective rising and setting, and what they did signifie : and whereas others used to assigne the form of some Animal or other, to each ten degrees ; they assigned one to each single degree, and so made their harsolations or conjectural predictions accordingly. For Example ; to the first degree of *Aries* they assigned the figure of a Man, holding a Side or hook in his right hand, and a Sling in his left ; to the second, a Man with a Dogshead, his right hand stretcht forth, and a staff in his left; and so of the rest : then annexing the signification to each, they determined, that he, who should have the first degree of *Aries* for his Horoscope, should be some part of his life a Husbandman, and the rest of it a Soldier; that he, who should be born under the second, should be contentious, quarrelsom, and enviou s; and so of the rest, all which *Scaliger* hath fully deduced from *Aben Ezra*. In a Word ; what ever knowledge either the *Egyptians* or *Chaldeans* had of the Stars ; certain it is, they referred it wholly to *Astronomantie*, or Divination by Stars : and therefore among them there flourisht, no true and genuine *Astronomy*, but a spurious and false, one *i. e. Astrology Divinatory*, or the fraudulent Art of Fortune-telling by the Heavens.

Berosus (whom we formerly mentioned) coming into Grece, a little after the death of *Alexander*, is discovered to have brought with him nothing sollid touching *Astronomy*, but only Judicial *Astrology* ; for which, as a thing new, and strange to the people, he was highly esteemed, as *Vitruvius* and *Pliny* remark. And *Eudoxus*, who had returned out of *Egypt* before that, well knew what sort of *Astrology* this was (the principal Contrivers and Founders of

lib. 9. cap. 7.
lib. 7. cap. 37.

of which are ſaid to have been *Petoſires, Necepſus, Eſculapius*) but he highly contenmed it as *Cicero* remembers, and brought home no other fruit of his tedious Travells, beſide a liſt of ſome *Eclipſes*, and the varieties of the motions of the wandering Stars, by which he firſt eſſaied to compoſe accommodate Hypotheſes, as we have formerly hinted. Nay, *Plato* himſelf, who was Companion to *Eudoxus*, for thirteen Years together, in *Egypt* : profeſt : that he could attain nothing ſollid and ſatisfactory touching thoſe Stars, and therefore placed all his hope only in the ſagacity and induſtry of the *Grecians*, ſuch as he knew *Eudoxus* to be. ” For, having firſt recounted what ever he knew concerning ” them ; he ſaith, It is to beleeved that the *Grecians* make ” more perfect whatſoever they receive from *Barbarians* ; and ” therefore is it fit, we allow the ſame, touching the argument ” of which we have diſcourſed. Truth is, it is difficult to ” find out the way, how all theſe Apparences, ſo involved in ob- ” ſcurity may be explicated : nevertheleſs there is great hope ” that things of that ſort will be better and more advantageoſ- ” ly handled, than they were delivered to us by *Barbarians*.

From the *Egyptians* and *Chaldeans*, therefore (as *Aſtronomy* her ſelf, while young and rude) we come to the *Græcians* : and the moſt antique record of Syderal Obſervations to be found among them, ſeems to be that of *Heſiod;* who in his Book of Weeks and Dayes teacheth Husbandmen the moſt opportune times of reaping, ſowing, and other labours of Agriculture, from the riſing and ſetting of the *Pleiades*, and *Hyades*, and *Arcturus*, the *Dog-ſtar*, and *Orion*:

Πληϊάδων Ἀτλαγενέων ὀπιτελλομενάων.

Donec Pleiades, quæ & Atlantiades, exoviuntur, &c.

And I cannot tell, whether it were that book, or ſome other, that *Pliny* meant, when ſpeaking of *Heſiod*, he ſayes, *Hujus quoꝗ, nomine extat Aſtrologia*, there is extant an *Aſtrology* of his. However, we are here to remark two things, in order to our more exact diſquiſition ; the *Firſt* is, that the Ancient *Greeks* principally attended to theſe riſings and ſettings, aſwell that they might diſtinguiſh the ſeveral Seaſons of the Year, as that they might fore-know Rain, Winds and other diſpoſitions of the Air, uſually attending thoſe Seaſons. And hereupon, *Thales, Anaximander, Democritus, Euctemon, Meton, Eudoxus*, and many

D 2

2. de divinat.

in Epinom.

Artid. 6.
*And after them
to the* Grecians
*among whom
the moſt anci-
ent mention of
Aſtron. is in*
Heſiod.

*Ex Gem Ptol.
& aliis.*

many others, composed certain *Parapegmata*, Tables, (as E-
phemerides, or Diaries) in which they inscribed each Day of
the Year, with the particular Stars rising or setting on each Day,
and what mutations of the Air each one did portend. Such a
Darapegme as these, was composed likewise by *Julius Cæsar*
himself, for the Horizon of *Rome* ; in allusion where-to he
might justly own, what *Lucan* said for him,

lib. 10.

> *Nec meus Eudoxi fastis superabitur Annus.*

lib. 1.

And, him doubtless, did *Ovid* translate into his *Fasti* ; pro-
mising in the beginning, that he would sing of the Stars and
Signes, that rose and again descended under the Earth. But, to
keep close to the *Grecians;* among them, he was held a great
Astrologer, who had discovered and observed only these risings
and settings here spoken of ; and so of whom that might be
spoken, which *Catullus* said of *Conon,*

> *Omnia qui magni dispexit lumina Mundi,*
> *Stellorumꝗ ortus comperit, atꝗ obitus.*

in diedus.

For, before the Advent of *Berosus,* this was the only [Ἐπιση-
μασία] Præsignification or Divination by the Stars, the *Grecians* had among them : unless what *Hesiod* hints, in his

> Πρῶτον ἔνη τετ'ς τ̄, κ̀ ἑβδόμη, ἱερὸν ἡμαρ,
> *Primùm prima dies, & quarta & septima sacra, &c.*

where he points out, what Dayes of the Moon were accounted
Lucky, and what Unlucky.

The *Second* observable is ; that among the *Grecians,* and indeed among divers other Nations, beyond all Memorials of
either Traditions or books, the Stars were reduced to certain
Images, or Constellations, and denominated accordingly (as
their names yet shews) as it pleased the fancies of Husbandmen,
Shepheards, Mariners and the like, who used to be vigillant and
gazing upon the Heavens in clear Nights. Though there have
been some Constellations added of latter times, as that of the
lesser Wain, by *Thales,* which *Lacrtius* and *Tatius* recite out of
Callimachus, who also took the same elswhere, and that of *Berenices*

lib. 1. de vit.
Isogode Com.
Berin.

renices Hair, removed into Heaven by *Conon,* as *Catullus* relates. *Cleoſtratus* likewiſe (as we have it from *Hyginus*) found *lib. 2. Aſtron.* out the *Kidds :* though, (which *Pliny* moreover attributes to *lib. 2. cap 8.* him,) his invention of the Signes in the Zodiack, is ſo to be un-derſtood, as that he taught men through what Signes the Sun and other *Planets* paſſed. But (that we may couch alſo upon this) at firſt, the *Grecians* had only Eleven Signes in their Zodiack ; and it was long after ere they came to add the twelfth, in imitation of the *Egyptians,* who (as may be collected from *Servius, in 1. Geog. l. 8. Marcianus,* and others) inſtead of the *Clawes* of the *Scorpion,* placed *Libra,* the place deſtined to *Auguſtus,* by *Virgil,*

-----*Ipſe tibi jam brachia contrahit ardens Scorpius.*---   *1. Georg.*

They added the Twelfth, we ſay, to the end, that as the whole Compaſs of the Zodiack was divided into *Dodecatemoria* (as they call them) twelve equal parts, ſo it might conſiſt alſo of twelve Signes. Albeit, being (as it were) neceſſitated to make uſe of ſuch Signes, as had been brought up, rather by chance, than Art ; thoſe 12. Signes were not exactly proportionate to the 12. Diviſions of the Zodiack, but took up more ſpace ſome, than others as in particular, *Leo* poſſeſt more room than *Can-cer ; Taurus* than *Gemini.* I ſay, than *Gemini,* which though compoſed of *Caſtor* and *Pollux ;* in ſo little ſpace as is allowed them, it is impoſſible the one ſhould riſe, when the other Sets, and both in the *Eaſt :* but this *Empiricus* interprets of the two *1. adven.* *Hemiſphears.* I omit to infiſt upon this, that all Nations had *Phyſic.* not the ſame Conſtellations : as among the *Egyptians* was no *Bear,* no *Cepheus,* no *Dragon;* but other formes or repreſentati-ons, as *Tatius* reports ; and ſhall add only, that *Eudoxus* ſeems to have been the firſt, who partly out of the *Egyptian* Fi-gures, partly out of the *Grecian,* furniſhed the whole Zodiack with Images reſembling the *Aſteriſmes,* (as men had fancied, at leaſt) and cauſed them to be drawn on a Globe, or ſolid Sphear. For, *Aratus* (upon whoſe Poem, intitul'd Φαινόμενα, *Apparen-ces,* there have been ſo many Commentaries ſet forth, as that no fewer than forty have been extant in Greek ; beſides thoſe of *Cicero, Germanicus, Avienus,* and other Latin Interpreters.) did no more, but only expreſs in verſe, what *Eudoxus* had ſaid before in proſe, of this argument ; as *Hipparchus Bythinus* de-

D 3   mon-

monſtrates. I know not, whether it would be ſeaſonable for me, here to advertiſe, that it is no wonder *Aratus* erred ſo groſly in many particulars; conſidering that (as is written in his life) he Living with *Antigonus Gonata,* in the quality of his Phyſician, and *Nicander* in the quality of his *Aſtrologer ;* and both were good at Poetry : *Antigonus* commanded the Phyſician to give him a tryall of his Poeſie, upon an Argument in *Aſtrology ;* and the *Aſtrologer* to give another of his, upon ſomthing in *Phy-ſick :* delivering to the one, the Book of *Eudoxus ;* and to the other, all that was extant of *Treacles, Antidotes,* or *Counterpoiſons.* So each wrote of what he did not well underſtand. One thing I ſhall not forget ; and that is, that the *Phenomena of Euclid,* who lived neer about the ſame time, and taught at *Alexandria*

(as in the Memorials of *Pappus*) were quite of another kind ; being indeed no other, but certain Principles of *Aſtronomy,* con-cerning the figure of the World, and the Circles of the Sphear, and chiefly, that of the Zodiack.

But, to return back to the more primitive *Greeks ;* I remember I ſaid, that *Thales Meleſius* was accounted the Firſt, who after old *Heſiod* and *Homers* Dayes, enquired into the Order of the Stars. And, certainly He was the Man, who among the *Grecians* may challenge the Palme ; as to Antiquity ; for,

Apuleius calls him, *ut antiquiſſimus, ſic peritiſſimus Aſtrorum Contemplator,* and *Eudemus* in *Laertius* atteſteth, that this was the Opinion of moſt, adding moreover, that *Xenophanes* and *Herodotus* highly admired him, for that he had firſt predicted the *Eclypſes* and Converſions of the Sun ; and that *Heraclitus* and *Democritus* witneſſeth as much. And whereas *Apuleius* further ſubjoyns, that he found out the motions and oblique tracts of the Syderal Lights : *Pliny* aſcribes that to *Auaximan-der,* a Diſciple of *Thales Mileſius,* (whence he was ſaid *Rerum fores aperuiſſe,* to have opened the Doors of Celeſtial matters) and *Diodorus* to one *Oenopides Chius :* which *Thales* could not yet be ignorant of the Obliquity of the Zodiack, when he had written of the *Solſtices,* and *Equinoxes,* and had converſed a long time with the *Egyptians* in their own Country, as *Laertius* re-members. Further, it is delivered to us, that among others, he predicted that notable *Eclipſe* of the Sun, which hapned in the time of the warre betwixt the *Meads* and *Lydians ;* which he could not doe by any other reaſon, but only becauſe, coming
                                                                      newly

newly out of *Egypt*, he had learned, that *Eclipſes* generally return upon the ſame Day after the ſpace of nineteen Years ; and
having taken notice of one, that fell out 19. Years before, he
concluded that there would be one at ſuch a time. Nor is there
reaſon why any ſhould think, that otherwiſe his whole life
might be ſufficient to obſerve all the motions of the Sun and
Moon, as from thence to be able to invent all things neceſſary for
the calculation of the times of their Several *Eclipſes.* Moreover, it *in vit. Dionyſ-*
doth not appear, how by any other way, but that *Helicon Cy-*
zicenus came afterward to fore-tell that *Eclipſe* of the Sun
(mentioned in *Plutarch*) for which he was ſo much admired by
Dyoniſius, and rewarded with a Talent of Gold. Nor likewiſe,
how *Sulpitius Gallus* could fore-tell that other of the Moon,
which as moſt opportunely predicted to the *Roman* Army, then
ready to joyne battell with the *Perſian,* is ſo higly celebrated,
not only by *Plutarch* and *Pliny,* but alſo by *Valerius, Quintilian,*
and other Hiſtorians : for other Rule for the calculation of future *Eclipſes,* there was none before *Hipparchus,* who invented Hypotheſes and Tables fit for that purpoſe. Beſides, what *La-*
ertius imputed to *Anaximander, Plinius* as confidently imputes
to one *Anaximenes,* an Auditor of his : (namely that he ſhould
be the Inventor of that *Gnomon,* by which the Converſions of
the Sun, or the Solſtices and Equinoxes, were indicated, and
that he ſet up ſuch a one at *Lacedemon.*) Neer upon the ſame *then of* Pytha-
time was it, that *Pythagoras* is ſaid to have firſt diſcourſed goras, *and his*
(though *Phavorinus,* in *Laertius,* confers that honour upon *Diſciples.*
Parmenides,) that *Lucifer* and *Veſper* was one and the ſame Star
of *Venus.* Now, whether may we conceive, that he borrowed this
of the *Egyptians,* from whom being taught, that not only *Venus,*
but *Mercury* alſo, were carried round about the Sun, as their
Center, ſo that one and the ſame might be both Morning and E-
vening Star : poſſibly, from thence he might take the hint of his
Conjecture, that the Sun was the Center of not only thoſe two,
but of the other Planets alſo, and conſequently of the whole
World : and moreover that the Earth it ſelf, as one of the *Planets,*
moved about the Sun? For truely, this was an eminent and conſtant Tenent in his School ; as may be underſtood not only from
Ariſtotle in the general, but alſo from *Laertius* in particular of *2. de coelo. 13.*
Philolaus, and from *Archimed* of *Ariſtarchus,* both *Pythagorus* *de Arenar,*
his Diſciples : that we may not rehearſe all thoſe many paſſage s *num.*

in

Philolaus,
Ariſtarchus,
Timæus.

de don. ad pæon.
in Timæun.

After theſe ſuc-
ceeded Cleo-
ſtratus.

Meton. &c.

in *Plutarch*, concerning this memorable particular ; nor name thoſe, who held, that the Earth was not ſo much moved about the Sun as dayly turned rouud upon an Axis of its own ; as *Timæus*, a *Pythagorian* alſo, who is therefore by *Syneſius* eſteemed, after *Plato*, the *moſt excellent Aſtronomer*.

Furthermore, in the next Age after *Thales*, or neere upon, comes *Cleoſtratus* (the ſame who was beleeved to have deprehended the Signes of the Zodiack) and he, ſeriouſly remarking that the Intercalation, which as we ſaid, was wont to be made every fourth Year, celebrated with the Olympick Games, did indeed reſtore the motion of the Sun to the ſame Day again ; but did not reſtore the motion of the Moon till the eight Year, or two Olympiades, in which the intercalatory Dayes amounted to ninety Dayes, or three months : He, we ſay, thereupon interduced, inſtead of the *Tetaēteris*, or ſpace of four Years, the *Oētaeteris*, or ſpace of eight Years, which compleatly paſt, the New-Moons, and Full-Moons would returne again on the ſame Dayes. But, when in ſhort time men had perceived, that this Inſtitution failed them, in exaētneſs of computation ; and that ſundry wayes had bin attempted to cure this uncertainty: at length riſeth up *Meton*, ſomwhat more ancient than *Eudoxus* ; and he demonſtrateth from the New-Moons, and Full-Moons Eclyptical, that they did not return upon the ſame Dayes, till after full nineteen Years : and thereupon he became the Author of the *Enneadecaeteris*, or *Period*, or *Cycle* of 19. Years. In reſpeēt of which diſcovery, together with the Heliotrope, or Sun Diall he made at Athens, and ſome other the like Inventions, he was in eminent eſteem among the *Athenians*. But as concerning that *Period* ; *Callippos*, familarly acquainted with *Ariſtotle*, diſcovering it to be too long, by the fourth part of a Day ; inferred, that from four Periods one whole Day ought to be detraēted : and ſo erected a new Period, or Cycle of Sixty ſix Years, or four times nine, at the end of which one Day was to be cut off; and this was called the *Callippik Period*, and remained in uſe for a long time together. After him ſucceeded *Hipparchus*, who deteēting this *Period*, to be yet too long ; demonſtrateth that after four *Callippik Periods*, or three hundred and four Years, there would remain one whole Day too much. And in truth, the experience of many ſucceding Ages declared, that to this detraētion of *Hipparchus*, nine or ten Years over and above were to be expeēted.

How-

However, it is worthy our notice, that the *Period* of *Meton*, together with the Coneꞔtion of it, applied by *Callippus*, was of long uſe in the Church under the name of the *Golden-Number*: though wanting the Application of *Hipparchus* his Correꞔtion: alſo, a miſtake of about four Dayes, relating to the New and Full Moons, crept into the account, even from the time of the *Nicene* Councel; which was one of the two main cauſes of the Reformation of the Kalender in the eighty ſecond Year of the laſt Age.

And now we have an opportunity to ſpeak more expreſſly of *Eudoxus*, ſo frequently mentioned. This man, well underſtanding, after his return out of *Egypt*, that not only the Sun and Moon, but alſo the five Errant Stars, did keep their courſes round in the Zodiack ; and ſo, as that aſwell the Sun and Moon, as thoſe wandering Stars did ſometimes vary their latitude, or deviate from the *Ecliptick* Line in the midle of the Zodiack ; (for, he thought that the Sun was alſo extravagant, as well as the reſt; and again, that the other Planets did not only go forward, but were alſo ſome times upon their retreat backward; and ſomtimes made a hault or ſtood ſtill : we ſay, pondering all theſe various motions in his mind, and caſting about what might be the reaſons thereof in nature; he at laſt imagined to himſelf, that beſides the *Aplanes* or Sphear of Fixt Stars, which being ſupreme, carried all the reſt toward the *Weſt*, there ought alſo to be allowed three other Sphears, aſwell to the Sun, as to the Moon, and four to each one of the other Errant Stars; of which one, and that the higheſt, ſhould follow the Impreſſion of the Fixt Stars, or rather of the *Primum Mobile* ; the next to that ſhould move counter to the Firſt, or toward the *Eaſt*; the third make the deviation from the *Ecliptick*, or midle of the Zodiack ; and the fourth, or loweſt, cauſe in the Stars their Direꞔtion, Station and Retrogradation, and that by a certain Vibration, or Waving to and agen. So that he ſuppoſed in all, twenty ſeven *Sphears*, and all thoſe Concentricall, that the Superior might carry on the Inferior, and theſe might be turned round within thoſe. Afterwards, *Callippus* adjoyned two *Sphears* to the Sun, two to the Moon, and one a peece to *Mars, Venus*, and *Mercury* : and ſo made thirty three. And *Ariſtotle*, to all the *Sphears*, which did not follow the motion of the *Aplanes*, or *Primum Mobile* (excepting only the Lunar *Sphears*) added as many more, which he called the Revolvent ones, to the end he might conform them to the mo-

Artid. 7. Eudoxus, *who firſt diſcovered the neceſſitie of manifold* Spheares.

E   tion

tion of the Inerrant *Sphear,* or *Primum Mobile* : and so in the whole he constituted Fifty six *Sphears* ; for as much, at least, as we can collect from his own context. Now all these, and even *Plato* himself likewise, thought that the Moon was the lowest of all the *Planets* ; next to her, the Sun ; and above the Sun the five wandering Stars: Nor indeed doth it appear, that *Archimedes* himself Living a whole Age or two after them, represented the *Planets* in any other, than this very order, in that so famous *Sphear* of his   In which though *Claudian* tells us, that no more was represented, but only the motions of the Sun and Moon ;

in Epigram.

> *Percurrit proprium mentitus Signifer annum,*
> *Et simulata novo Cynthia mense redit :*

2. de divinat.

       Yet *Cicero* adds other motions, when speaking of *Archimedes,*
„ he saith ; when he collected together the motions of the
„ Moon, Sun, and five wandering Stars ; he did the same as that
„ God, who in *Platoes Timæus* framed the World, that one and
„ the same Conversion might regulate sundry motions, most dif-
„ ferent each from other in slowness and swiftness . But, *Hipparchus* afterwards finding, that aswell the Sun, as the Moon and the other five Stars did come somtimes neerer to the Earth, and sometimes again mounted up farther from it; and plainly perceiving that that particular apparence could not possibly be explicated by those *Sphears,* that were all Concentrical to the Earth : therefore, wholly rejecting them, he resolved, that the motions of the *Planets* were to be accounted Eccentrick ; and though he could not himself determine each particular, he yet demonstrated the way, in which *Ptolomy* afterwards insisting, accomplisht the Invention. But, before wee advance further, we are to commemorate two or three Persons of note, by whose Observations both *Hipparchus* and *Ptolomy* profited very much. One was *Timocharis,* who, about three hundred Years before Christ, among other things relating to the Fixt Stars, observed that that Star which is called *Spica Virginis,* doth antecede the point of the *Atumnal Equinox,* by eight degrees. And with him are we to joyn *Aristillus,* whose Observations of something about the Fixt Star; *Ptolomy* made great use of, in order to his demonstrating that the Fixt Stars never change their latitude. Afterwards (scarce

in Age) ſucceded *Eratoſthenes*, who being *Library-keeper* to *Ptolomy Evergeta* the fomer, perſwaded him to ſet up the *Armillæ*
in the *Porticus* of *Alexandria;* which *Hipparchus* and *Ptolomy*
afterwards made uſe of ; and himſelf, among other things obſerved, that the Obliquity of the Zodiack was of twenty three degrees, and fifty one minutes; which account *Hipparchus* and *Ptolomy* conſtantly adhered to.

Now that we may at length remember the great *Hipparchus*,
who floriſh't neer upon an hundred and forty Years before
Chriſt : truely, we find it no eaſy task to recount, how highly
Aſtronomy was beholding to him. For, in the firſt place, examining that foreſaid Obſervation of *Timocharis*, with ſome o-
thers; albeit he could not conceive them to be in all points exact,
yet becauſe himſelf had found that *Spica Virginis* did not antecede the Equinoctial point by more than ſix degrees, and the
other Stars in the like Proportion : he thence underſtood, that
the Fixt Stars alſo were moved Eaſtward according to the Zodiack ; and thereupon wrote a Book of the *Tranſgreſſion of the
Solſtices & Equinoxes.* And, being that in his time, as not
long agoe in *Tycho Brahes*, there appeared a certain New Star,
,, he therefore came to doubt (to ſpeak the language of *Pliny*
,, concerning him) whether the like happened often, or not ; and
,, whether thoſe Stars, that were thought to be fixt, had alſo ſome
,, certain motion peculiar to themſelves. Wherefore (as the ſame
,, *Pliny* goes on) he attempted a task of difficulty ſufficient even
,, for the Gods themſelves, namely to number the Stars for Poſte-
,, rity, and reduce the heavenly Lights to a rule, ſo that by the help
,, of Inſtruments invented, the particular place of each one, together
,, with its magnitude, might be exactly deſigned : and where-
,, by men might diſcern, not only whether they diſappeared, or
,, newly appeared, but alſo whether they removed their Stations;
,, as likewiſe, whether their magnitude encreaſed, or diminiſhed ;
,, Leaving Heaven for an Inheritance for the Witts of ſucceding
,, Ages, if any were found acute and induſtrious enough to com-
,, prehended the myſterious orders thereof. And this was the
firſt time when the places of the Fixt Stars were obſerved and
markt out according to Longitude and Latitude : and that Catalogue of the Fixt Stars, which he compoſed, is the very ſame,
which *Ptolomy* afterward inſerted into his *Almageſt.* In the
next place, he denoted was poſitions ſundry Stars had in reſpect

E 2   each

*Ptolom. lib. 1.
cap. 11.*

Artid. 8.
Hipparchus,
*who firſt obſer-
ved the places
of the Fixt
Stars, accord-
ing to Long.
and Latitude.*

lib. 2. cap. 25.

lib. 7. cap.

each of other ; whether they were posited in a right Line ; or in a triangular form ; or in quadrate or square, &c. as is manifest even from *Ptolomy* himself. Further, though the motions of Sun and Moon were already in some measure known ; he yet made that knowledge much more exact. For, He did not only much correct the *Callippick Period,* formerly spoken of, but also, having collected a long Series of *Eclipses* (namely, from the time of those *Babylonish* ones, in the Dayes of *Mardocempades,* down to those observed by himself, for full six hundred Years together,) and remarking, that neither the like *Eclipses* did return on the same Dayes, after the space of every nineteen Years, nor that after some recurses of ten *Novennales,* or ten times nine Years, any such *Eclipses* happened at the times supposed ; and that the cause thereof consisted both in the various Latitude of the Moon, and the anticipation of her *Nodi,* or Knotts, and her Eccentricicy, by reason whereof her motions to her *Apogeium* were found to be sometimes slower, and those to her *Perigeium* more speedy : therefore, we say, He comprehended and gave Reasons for all these difficulties, and composed certain Hypotheses, and according to them, certain Tables, by which he could safely and exactly calculate and predict what *Eclipses* were to follow, how great they were, and when. And this was it, which *Pliny* remembred, when having spoken of *Thales,* and *Sulpitius Gallus,* he comes to mention *Hipparchus.* After ,, these (saith He) *Hipparchus* foretold the courses of both Lu-,, minaries, for six hundred Years to come ; comprehending the ,, months, Dayes and hours of Nations, and the Scituations of ,, Places, and turns of People : his age testifying, that he did all ,, these great things, only as he was partaker of Natures Councels. For, it must be, that *Hipparchus,* besides the precise times, when such or such *Eclipses* were to be visible to the Horizon of *Rhodes,* or *Alexandria,* pointed forth also some Countries, and principal Citties, together with the Designation of the Months in use among them; as also the very Days and hours when each *Eclipse* would happen; and other prædictions succeding to *Rome,* in the Dayes of *Pliny.*

Again, it is well worthy our recital, that *Hipparchus* labouring with long desire both to constitute *Hypotheses,* and reduce into Tables the motions of the other *Planets,* or five wandering Stars ; and yet not being able to furnish himself either from the *Egyptians,* or from his Country men the *Grecians,* with any

com-

competent Obſervations reſpective to thoſe *Planets*, (for while
the places of the Fixt Stars remained unknown, it was impoſſible any ſuch could be made) and again thoſe he had himſelf
made, were of a much ſhorter time, than was requiſite for the
eſtabliſhing any thing certain and permanent in that ſort : He
therefore only digeſted ſuch Obſervations as he had recorded by
him, into the beſt order and method he could deviſe ; and ſo
left them for their uſe and improvement, who ſhould come after
him, in caſe any were found capable of underſtanding and advancing them. And at length, by good fortune, it ſo fell out,
that thoſe his Obſervations came into the hands of *Ptolomy* ;
who comparing them with his own, and finding them judicious
and exact, thereupon firſt began to erect both *Hypotheſes*, and
Tables of Motions fit for thoſe *Planets* : yet not without much
timerouſneſs and diffidence ; becauſe his Obſervations being but
few, nor of ſufficient time, he durſt not promiſe himſelf any certainty of his Tables for any conſiderable ſpace, or number of
Years. But, for more aſſurance let us hear his own ingenious
Confeſſion in that point. The Time (ſaith He) from whence   *Almageſt. lib. 9*
„ we have the Obſervations of the *Planets* ſet down, is ſo vaſtly   *cap 2.*
„ ſhort, in compariſon of the greatneſs of Coeleſtial viciſſitudes,
„ as that it renders all predictions, that are for any great number
„ of Years to come, infirm and uncertain. And therefore I
„ judge that *Hipparchus* (that zealous lover of truth) conſidering
„ this difficulty, and withall receiving not ſo many true Obſer-
„ tions from the Ancients, as he bequeath'd to us, undertook
„ indeed the buſineſs of the Sun and Moon, and demonſtrated
„ that it might be performed, by equal and circular motions: yet,
„ as for that of the *Planets*, thoſe Commentaries of his, which
„ have come into our hands, clearly ſhews, that he attempted it
„ not : but collecting all his own Obſervations concerning them
„ together, into one order and method, for their more commodi-
„ ous uſe, reſigned them to the induſtry of after times; having firſt
„ demonſtrated, that they were not congruous to thoſe *Hypothe-*
„ *ſes*, which the *Mathematicians* of thoſe Dayes made uſe of.
„ And, for Others ; ſure I am, that either they demonſtrated
„ nothing at all, or elſe only attempted the buſineſs, and left it un-
„ finiſht. But, *Hipparchus* being eminently knowing in all
„ kinds of learning, conceived, that he ought not (as others had
„ done before him) to attempt, what he ſhould not be able to

E 3   ac-

accomplish. So that we see, *Ptolomy* was the first, who from true Observations, reduced the Motions of the *Planets* into *Hypotheses* and Tables correspondent.

Artid. 9.

Betwixt Hip-parchus *and* Ptolomy, *came* Sosigenes, *of* Alexandria, *by whose help* Jul. Cæsar *endeavored the reformation of the* Calandar.

But before we speak more particularly of him, who lived about an hundred and thirty Years after Christ ; forasmuch as in the space of time betwixt *Hipparchus* and *Ptolomy*, these studies so florisht at *Alexandria*, as that *Julius Cæsar* returning thence, brought along with him that *Sosigenes*, by whose assistance he endeavoured the restitution of the *Calendar*, and so may be thought to have propagated the Study of Astronomy among the *Romans* : let us reflect a little upon that time, and see what care they then had of Celestial matters. In the first place, we are to lay aside the Commemoration of *Sulpitius Gallus* (of whom more then once afore, as one that falls not under this account, concerning whom we may not yet forget, what *Cato* is induced by *Cicero*, saying. While we saw that *Gallus* dye, that familiar friend of thy Father, *O Scipio*, who was restless in measuring Heaven and Earth; I say, while we saw him dying even in that Study. How often did Day oppress him, when he had set himself to observe and describe somthing in the Night? and how often did Night oppress him, when he had begun his Speculations in the Morn? How was he delighted, when he had a long time before predicted to us *Eclipses* of the Sun and Moon? &c. For he was a man dearly singular, and in an Age when so great ignorance and neglect of good Arts tyrannized over mens minds, being himself studious and inquisitive, could not but have borrowed his skill either from *Egypt* or *Greece*, where having obtained a *Series* of *Eclipses*, and the way of deducing them through the circuit of of nineteen Years (as we said afore) he became able to calculate them, so as *Cicero* relates   For, as to the rest ; how great doe you think was the ignorance and neglect, nay even contempt of studies of this nature among the *Romans*? Why, truely so great, as that *Virgil* could not dissemble it, in the Poesy attributed to *Anchisa*, according to which the *Romans* should indeed come to rule the World ; but yet should yeeld to others, in learning to know the Stars, and describe the Heavens.

- - - - - - - - - - - *Caeliꝗ meatus*
Describent radio, & surgentia sidera dicent.

6. *Æml.*

And *Cato* himself is cited by *Agellius* to have left in writing
that

that it was not lawfull to write what is in a Table kept by the High Preiſt, how often ſcarcity of Proviſion would happen, how often the light of the Sun, or Moon ſhould be darkned : ſo far, ſaith *Gellius*, did *Cato* contemne the Science of *Aſtronomy*, and thought it uſeleſs either to know, or fore-tell the *Eclipſes* of Sun and Moon. Furthermore, though from times as high as *Numa*, the *Romans* made ſeveral Intercalations; yet they took all their art of that Sort, from the *Greeks* : and *Pliny* remarks, that in *France*, *Spain* and *Africa*, there was no one man, who could ſo much as tell the Riſing of the Stars. Nevertheleſs we are not to forget, that among the *Gauls* was one *Pythias*, the *Philoſopher* (as *Cleomedes* calls him) of *Maſilia*, who about the time of *Alexander* of *Macedon*, found the proportion of the *Gnomon* to the Solſticial ſhadow, to be the ſame at *Maſſilia*, as *Strabo* tells us *Hipparchus* had obſerved it at *Byzantium* ; who firſt attempted the *Northern Ocean*, and diſcovered the utmoſt *Thule*, in which *Cleomedes* cohærently proves the Summer Tropick to be the ſame with the *Polar*, or greateſt of all wayes apparent ones: and who (as from his Book, *de Oceano*, may be inferred) was excedingly curious to find out what was the Poſition of Heaven, reſpective to the variety of Countries and Climates. But, not ſo ſoon to digreſs from the *Romans* ; *Pliny* delivers, that in thoſe firſt times of *Romes* being a Common-wealth, the Invention of Dialls was very raw and imperfect : for that they had only the riſings and ſettings named, out of twelve Tables. That after ſome Years, they added the Meridian, and by the indication of a certaine columne, the laſt hour : nor that neither, but only in clear weather, even as long as till the firſt *Punick* warre. Afterward they advanced ſo far, as to make one, or two Sun Dialls ; but not with lines exactly correſpondent to the hours, untill about an Age after, when *Q. Marcus Philippus* ordered the buſineſs more diligently and ſuccesfully. And, becauſe the hours of the Day remained yet uncertain in dark and cloudy weather, *Naſica Scipio* began to divide the Hours of Day and Night equally by Water diſtilling from Veſſel to Veſſel, and called it *The Diall within Doores*, in the Year *Urbis conditæ* DXCV. And till then, ſaith *Pliny*, *Populi Romani indiſcreta lux fuit*. And thus much of *Hipparchus*, and ſome *Aſtronomers* betwixt him and the Prince of them all, *Ptolomy*.

And of him, ſo great is his name, all we need to ſay, is only, that

lib. 2. c 28.

1. Meteor 7. Pythias Maſſilienſis, *a Gaul, and contemporary to* Alex. *of* Maced.

lib. 7. cap. 60.

Quintus Marcius Philippus, *&c.*

Naſica Scipio, Romans.

Artic. 10.

Ptolomy, *the true Founder of* Aſtronomy *in one intire ſtructure.*

that He was the *very Founder of the Art, or Science of Aſtronomy.* For, though *Hipparchus* had indeed, as it were hewn out the Stones and Beams fit for ſo noble a Structure, and prepared good part of the Materials ; yet was it *Ptolomy* alone who put them into Order and Form, and by adding many admirable Inventions of his own, by infinite labour and coſt, erected that ſo famous Building, worthily called Μεγάλη Σμύταξις, *the great Co-ordination, Conſtruction, or Compoſition ;* which conſiſting of no leſs than thirteen Books, contains all the Doctrine, that could then be advanced, concerning the Sun, Moon, and aſwell the Fixt, as wandering Stars. And, albeit one Day teacheth another, and that (as Himſelf had truely foretold) there came others after him, who ſaw good cauſe for the Caſtigation and Correction of many things delivered in that Work : yet, in the general, the Art he had inſtituted, remained firm and conſtant, and was afterwards imbraced, not only by the *Alexandrians,* but alſo by all the *Arabians,* Latins and others, who devoted themſelves to the ſervice of *Urania,* ever ſince. For, that the Study of her Celeſtial myſteries continued in great eſteem and Veneration, at *Alexandria,* for ſome Ages after his deſeaſe, may be undeniably atteſted, not only from hence, that (among others) both *Theon,* and *Pappus,* named *Alexandrians,* were eminent therein ; of which the one put forth eminent Commentaries upon *Ptolomies* Works, and the other, among ſundry excellent peeces, of which his ſixth Book of *Mathematicall Collections* is one, obſerved, that about four hundred Years from Chriſt, the *Obliquity* of the *Ecliptick* was not ſo great, *as Erathoſthenes, Hipparchus* and *Ptolomy* had conceived : but neer upon the ſame we diſcover it to be in our Dayes : we ſay, that this is not the only Monument that is extant of the flouriſhing of *Aſtronomy* at *Alexandria,* long after *Ptolomy* had given it ſo great a Reputation there ; but there remains another as freſh and lively, which is the memorialls of thoſe *Patriarchs* of the *Alexandrine Church,* to whoſe judgement the determination of that great diſpute about the true time of Eaſter, was thought fit to be wholly referred, aſwell by the *Nicene Councel,* as by divers learned *Biſhops* afterward, and by Holy *Leo* himſelf, then Pope. Now, among theſe *Patriarchs* were *Theophilus, Cyrillus and Proterius,* whoſe advice and directions were thought neceſſary, in regard that the Controverſies raiſed about the Celebration of *Eaſter,* about the time of the *Veneral* Equi-

Who next reſigned it to Theon *and* Pappus, *both* Alexandrians.

Equinox, about the Full-Moon next following, and about conſti-
tuting certain conſtant Rules reſpective to them ; could not be
better compoſed, than by the definitive ſentence of theſe Pre-
lates, who Living at *Alexandria,* where *Aſtronomy* was in ſuch
Height, had the advantage of others, in point of knowing thoſe
things which were requiſite to the finding out of the truth. But,
of the *Arabians,* who in the Study of *Aſtronomy* ſucceeded the
Alexandrians, and tranſlated into their own language, the *Great
Compoſition* of *Ptolomy,* which they called *Almageſtum* ; the
Firſt, and moſt worthy to be remembred, was *Albategnius* other- *and long after,*
wiſe called *Mahometes Aractenſis,* born of a Family of the *to Albategnius,*
Dynaſtæ of *Syria* ; He about 800. Years after Chriſt, made *then to*
divers Celeſtial Obſervations, partly at *Aracta,* and partly at
Antioch : and found, both that the Apogeium of the Sun, ſince
the Dayes of *Ptolomy,* was advanced to the following Signes ;
and that the Stars did regreſs toward the Eaſt, one degree, not
in the ſpace of a hundred Years, as *Ptolomy* alſo had deſigned ;
but of ſomwhat leſs than ſeventy ; as alſo, that the Obliquity of
the *Ecliptick,* according to *Pappus* his Theory, was leſs (*viz.*
above 23. degr. 35. minutes) with many other particulars concerning aſwell the Fixt Stars, as the Planets : whereupon he both
corrected *Ptolomy* in many things, and compoſed new Tables,
and wrote a Book intituled, *De Scientia Stellarum.* After him, *Alphraganus*
within 2 or 3 Ages following, ſucceeded *Alphraganus, Arzachel,* *and other Ara-*
Almeon, and other *Arbians* ; among whom, (as being already *bians.*
tainted with that ſuperſtition which had corrupted the ſimplici-
ty of *Aſtronomy,* with *Aſtrological* Fooleries) ſome certain *Jewes,*
as ambitiouſly affecting the glory of Divination as the others, intermixed themſelves.

  After them, for a long time, the Worſhip of *Urania* lay neg- *Artid.* 11.
lected, nor did *Aſtronomy* receive any the leaſt (conſiderable) *Alphonſus, K*
advantage by Obſervators ; till neer about four hundred Years *of Caſtile, who*
ſince, *Alphonſus* King of *Caſtile* and *Lion,* being himſelf alſo *made and named*
toucht with the curioſity of *Aſtrological* predictions, and diſco- *the* Alphonſine
vering that the Tables aſwell of *Ptolomy,* as *Albategnius* were *Tables.*
not exactly agreeable with the Celeſtial motions ; ſet himſelf to
the compoſing of new ones ; and to that purpoſe convocated as
many *Arabians* and *Jewes,* as were eminent in thoſe Dayes for *A-*
ſtronomy; imploying them about Obſervations neceſſary to ſo great
a Work, and comparing with them thoſe of their Predeceſſors,

                    F   that

that ſo they might be the more exact in the performance of their task propoſed. And very memorable it is, that (as hath been credibly reported) He ſpent four hundred thouſand peeces of Gold on that undertaking : a munificence truely worthy the Heroick mind of ſo great a Prince, and which well deſerves to be had in perpetual commemoration by all lovers of Learning : but ſomwhat unhappily imployed, in reſpect the Perſons ſet a Work were not ſo ſtrict in ſtudiouſly and conſtantly obſerving, as ſcrupuloſly computing, directing their calculations not ſo much to what themſelves and others had really obſerved, as to certain traditional, myſteries, or Caballiſtical dreams : that we may paſs by their heedleſneſs, which *Regiomontanus* detecting, perceived, that they had miſtaken the true places of the Fixt Stars, by very neer two whole degrees ; as accounting the numbers of *Ptolomy,* as if they had bin conſtituted by him from the beginning of the Years of Chriſt. Which conſidered, we have the leſs reaſon to wonder, if the Tables compoſed by them, called from the Kings name, the *Alphonſine,* and ſometimes from the place, where they were made the *Toletane Tables* (whence alſo He, who was Preſident of that aſſembly of *Aſtronomers,* is ſaid to have been one *Iſaac Chanter* of the *Toletun* Synagogue) have been found, ever ſince the time of K. *Alphonſus,* to diſagree with the Heavens, and to require the review and caſtigation of ſome new and more

after whom, the Science lay neglected till, Georg. Peurbacchius *and* Ioh. Regiomontanus *aroſe, and again cultivated the ſame.*

faithfull hands. Thence forward *Aſtronomy* lay neglected, and almoſt buried in oblivion, (only *Thebitius* an *Arabian,* and *Prophalius* a *Jew,* obſerved in the mean time ſome ſmall matters, about the motion of the Fixt Stars, and the obliquity of the *Ecliptick,*) untill about two hundred Years ſince, *Georgius Peurbacchius,* and *Joh. Regiomontanus,* his diſciple, ſeemed to revive it. For, theſe worthy men delivered it out of the double cloud of ignorance and vanity, which the *Arabians* and *Jewes* had raiſed, to the Obſervation of its luſtre ; and kindled the Light thereof afreſh in *Germany* : reducing *Ptolomy,* providing Inſtruments, and making not a few faithfull Obſervations: though they were not ſo happy, as to bring their deſigne to that perfection they hoped and had propoſed to themſelves ; both of them dying in the middle and flower of their Age.

Articl. 12. *Then followed the moſt accute* Nich. Coper-

    Animated by their example, *Nicholaus Copernicus* (a *Boruſſian* born, and *Canon* of the *Cathedral Church of Warmes,* ſcituate neer *Fruemburgh,* in the ſame Country) about the beginning of the laſt Age,

Age, ferioufly addreffed himfelf to the Illuftration of *Aftronomy;* and reviving the long neglected Syfteme of the World excogitated by *Pythagoras,* he made many good Obfervations, in order to the compofing of new Tables But, forafmuch as he could not determine any thing concerning the Fixt Stars, befides their Promotion Eaftward, which they appeared to have made fince *Ptolomy's* time ; he therefore compofed fome *Canons* of their motions, and thofe as exact as poffibly he could : yet both thofe, and the *Prutenick Tables* that were built upon them, were incorrefpondent to the motions of the Heavens, though lefs incorrefpondent than the *Alphonfine.* Neverthelefs, the man is to be highly commended, both for his fublime perfpicuity, and modefty, in that forefeeing his Canons would need correction, he was wont frequently to exhort and encourage that ingenious young man, *Georgius Joachimus Rheticus,* deeply enamoured of the beauties of *Aftronomy,* to apply himfelf principally to the Reftitution of the Fixt Stars, and cheifly of thofe, which were in the Zodiack, or neer it, and with which the Planets might be moft conveniently compared : becaufe, without their reftitution, it was impoffible either to attain to the true places of the Planets, or to atcheive any thing of Moment or certainty, toward the advance of *Aftronomy.*

nicus, *who revived the doctrine of* Pythagoras, *concerning the Earths motion.*

And then at laft enters that Noble *Dane, Tycho Brahe,* upon the Theatre of *Aftronomy.* Who, as by in the impulfe of his Genius, being addicted to beholding and noting the Stars, even almoft as foon as he faw the light of them ; was fo much the more fpurred on by that advice of *Copernicus,* publifhed in the Works of *George Joachim* newly mention'd, by how much the more clearly he difcerned the impoffibility of determining the true and proper place of that famous New Star, (appearing in the Conftellation of *Caffiopeia,* from the beginning of November, in the Year M. D. LXXII. for above fixteen months together) without the reftitution of the Fixt Stars to theirs. For, He plainly perceived, that moft, if not all the Errors, which had bin found in *Aftronomy* even from its firft foundation or original, took their rife chiefly from hence, that the Fixt Stars really were not in thofe places, in which they were fuppofed to be, by Obfervators ; But fome of them were much neerer, and others again as much farther off ; and this, whether becaufe *Hipparchus* in the beginning had not with due exactnefs configned all the Fixt Stars

Articl. 13. *And laft of All, the noble* Tycho Brahe, *who out-did all the reft in difcoveries and inventions.*

F 2   to

to peculiar places, which indeed he had defigned by the Sextant of degrees (and truely it is very difficult at once to invent any thing of Moment, and perfect the invention) or whether becaufe the Tranfcribers of *Ptolomy,* out of carelefnefs, or ignorance, had corrupted the Original Text in many places ; or whether the additions afterward made, in refpect of the Stars progrefs to the following Signes, had occafioned any miftake and imperfection in that Theory ; or whether by any other unhappy caufe whatever. Now, in Order to this great Work, of rectifying thofe fundamental Errors, it pleafed Fate, that about the very fame time, that truely generous and never enough commendable Prince *William Landgrave* of *Haſſia,* had zealoufly devoted his mind and induftry to the fame care, of reftoring the Fixt Stars to their true manfions : but yet the honour he aimed at, was decreed only for the incomparable *Tycho* ; who in an Heriocal bravery of Soul, had now refolved with himfelf to enterpife no lefs than the Inftauration of the whole Science of *Aſtronomy* from its very fundamentals ; and fo to fpare neither labour, nor coft (efpecially while he was fo happy, as to have good part of his expences defraied by the liberal contributions of that eminent *Mecenas, Frederick* the *II.* King of *Denmark,* who thereby recorded his name in immortal Characters on the leaves of Fame) that fhould be neceffary to the making all Sorts of Obfervations requifite. As foon therefore as he had furnifhed himfelf with that *Aſtronomical* Colledge, or Tower for Obfervations, built by him in the *Iſland* of *Huenna,* to that purpofe affigned him by the King, and furnifhed that Heavenly Cittadel by him called *Uraniburg,* with ftore of exquifite and magnificent Inftruments Mathematical, he begun (having provided himfelf of fundry learned and competent Coadjutors) exactly to obferve the Altitude of the Pole, in that place, by the Circum-polary Stars. By which underftanding likewife the Altitude of the *Equator,* he pointed out the Equinoctial points, by the paffing of the Sun through them : and attending befides to the middle parts of *Taurus* and *Leo,* he found out the *Apogeium* of the Sun, and the Eccentricity of it, and deduced its Courfe from the point of the *Vernal Equinox.* Moreover, from *Venus,* in the Day time compared with the Sun, and in the Night with the Fixt Stars ; he endeavoured to fearch out the right Afcenfions, and Declinations of the Fixt Stars : which the Ancients had per-
for-

formed but fallacioufly, by uſing the Moon, not *Venus*, to that purpoſe. And his fucceſs was as exquiſite as his care in this, that he conſtituted that bright Star which is in the top of *Aries*, and ranged the chief of thoſe in order along the Zodiack : and then advancing to enquire or rather find out the diſtances of the reſt aſwell from them, as each from other, he defined both the right Aſcenſions and Declinations of all ; preſcribed their ſeveral Longitudes and Latitudes, and added to the Catalogue of the Ancients about 200. other Stars, wholly by them omitted. Becauſe the Ancients, Living in an Horizon much more Southern, had obſerved and ſet down neer upon 200. Stars, that are inviſible in the *Daniſh* Horizon, which is highly Northern : and *Tycho* again collected about 200. more than they could diſcern; and as being ſomwhat ſmall, he intermixed them among others of greater magnitude. Further, having in the mean ſpace, alwaies obſerved the paſſings of all the Planets through the Meridian, and their ſeveral diſtances from the cheif Fixt Stars neereſt to them : he laid ſuch ſollid foundations, as by them might be exactly known not only the true places of each, but alſo their ſeveral Motions. So that he came very neer the heighth of his noble hopes of building the whole Theory of *Aſtronomy* anew from the very ground, and of erecting compleat and everlaſting Tables for Calculation thereupon : but, alas! prevented by an immature death, He could not accompliſh his deſigne. It was very much, however, that He went ſo far, as to have recorded and bequeathed to Poſterity ſuch excellent Obſervations, by which *Kepler* was ſoon after enabled to compoſe an intire Theory, and make the Tables called the *Rudolphine* ; and by which, and others afterward contriveable, whatever can be deſired in theſe Tables, may be fully ſupplied and perfected. And this among the reſt deſerves ſingular commendations, that He left us the Fixt Stars re-inſtalled in their true manſions : wherein He alone, in few Years practice, performed and finiſhed that prodigiouſly great Work, which no man, from the Dayes of *Hipparchus*, had either attempted, or in any meaſure advanced.

I paſs by many other admirable diſcoveries of his ; as that he was the firſt, who demonſtrated all Comets to be carried freely through the Etherial Spaces ; that Refractions ought to be carefully conſidered and allowed for, and how ; that he perceived that the *Latitude* of the Moon ought to be augmented by more

than

than a Quadrant, or fourth part, than had been conceived ; that
He almoſt demonſtrately convinced the Latitudes of the Fixt
Stars to be varied ; that he excogitated an *Hypotheſis*, which all
thoſe, who cannot allow of the *Ptolomaicall*, or fear to allow
the *Copernican*, may well adhere to and defend ; with many
other things, as difficult in their Invention, as excellent in their
uſe. And obſerve only how vaſtly he tranſcended all that went
before him, in point of exactneſs and certainty. As for *Inſtruments Mathematicall*, it is well known, He made ſuch, as for
the condition of their matter, for the Vaſtneſs of their magnitude,
for the variety of forms, for the care of their elaboration, for the
preciſeneſs of their diviſions, and for the facility in uſing ; as the
World had never the like before. Again, ſo prodigious was
his and his Coadjutors ſubtility, diligence, induſtry ; that whereas
the Obſervations of *Hipparchus, Ptolomy,* and all others before
him, had bin marked out only by the Sixth or at moſt by the
twelfth parth of degrees ; he deſigned all his by the ſixtieth
parts of degrees, called Minutes, or Scruples, and very often alſo
by ſubdiviſions of Minutes. So that we may well demand what
compariſon can be made betwixt that groſs way found out by
Eraſtothenes, and approved and followed afterward by *Hipparchus* and *Ptolomy,* for the Obſervation of the Obliquity of the
Zodiack ; and that moſt fine and exact one invented by *Tycho* ?
His being, by a diviſion of the Meridian into 83. parts, and the
Interval of the Tropicks deprehended to take up 11. of them, it
appeared that the diſtance of one Tropick from the *Equator,* a-
mounted to 5. of thoſe parts and an half, or, by a reduction of them
again to degrees, of 23. degr. 51. Min. and $\frac{1}{3}$: and theirs, being
by an hollowed Hemiſpear of Stone, with a Gnomon erected in
the middle, as we have formerly deſcribed it ; and to what degree of ſubtility and exactneſs this way of commenſuration could
arrive, the meaneſt Novice in *Aſtronomy* may ſoon judge. That
Quadrant likewiſe of *Ptolomy,* ſo much admired by ancient Authors ; Pray, How vaſtly ſhort did it come of the perfection of
the leaſt that *Tycho* uſed? And the ſame may be ſaid of his *Rules*;
for, that thoſe *Armillæ,* ſet up by *Ptolomy* in the entrance of *A-
lexandria* had any thing in them comparable to thoſe erected by
Tycho, in his *Uraniburg,* cannot in the leſt meaſure be argued
from the other Inſtruments then in uſe. It is not neceſſary, we
ſhould here again review thoſe machinaments, or engines, which
the

the old *Egyptians* and *Babylonians* made uſe of, either in diſcerning the Signes of the Zodiack, or taking the Diameter of the Sun: or thoſe, which *Ariſtarchus* and *Archimedes* uſed, for commenſurating the ſame Diamater. Only we cannot but wonder, by the by, how *Ariſtarchus*, having aimed ſo neer the white of truth, in the matter of the Suns Diameter, and determining it to be the 720th. part of the Circle, or half a degree ; as is delivered by *Archimed* : ſhould yet err ſo widely in his Book of *Magnitudes* and *Diſtances*, as to make the Diameter of the Moon (which in truth; is very neer as great as that of the Sun) to be the 180th. part of the Circle, or 2. degr. when he called it *the Fifteenth part of a Signe* ; which miſtake of his was long ſince taken notice of by *Pappus*. Nor is there any neceſſity, why we ſhould ſurvey thoſe Inſtruments, that *Albateginus, Peurbacchius, Regiomontanus, Copernicus,* and other more moderne *Aſtronomers* uſed : conſidering, that beſides the *Rules* made by *Regiomontanus* (which *Bernardus Waltherus,* his diſciple, preſerved, and had recourſe to, in his Obſervations of the Suns Altitude) they came ſo ſhort of the leaſt of *Tychoes,* in point of exact reaſoning, and amplitude, that they deſerve rather to be perpetually forgotten, than remembred to competition. However it is ſeriouſly to be wiſhed, that the Obſervations made by thoſe incomparable Inſtruments of His, may ly no longer concealed from the World (for by ſingular Providence, they have been hitherto preſerved, as *Gaſſendus* atteſteth, *in the Life of Tycho*) but ſoon be brought to Light. And this aſwell for ſundry weighty Conſiderations there alleadged by *Gaſſendus* ; as for this, that not all the Stars, of which *Tycho* hath given a copious Catalogue, in his *Progymnaſmata,* may be found reduced to congruous Calculation (in as much as they doe not exactly correſpond with the Heavens) and that various Catalogues have been pretended from the ſame, which are very much different each from other: for all the difficulties hereupon depending may ſoon be removed, and all miſtakes rectified, by having recourſe to the Fountain, or Original obſervations, which will dearly declare, what hath bin already corruptly deduced, and what may be at length carefully and demonſtratively deduced from them.

    And, in the mean while, if *Hipparchus* his memory be ſo highly and (indeed) juſtly precious among learned men, for his great merrits in excogitating and framing Inſtruments, whereby to take the dimenſions, diſtances, motions &c. of Heavenly bodies :

de Arenar. num.

dies :

dies : certainly, that of our *Tycho* ought to be as highly efteem-
ed by us and all Pofterity ; fince he alone, for fo many Ages together was found, that durft not only imitate him in thofe fublime inventions; but fo imitate, as very much to exceed him. For
my part, truely ; fince *Hipparchus* may rightfully be called *Atlas the Second* : I fhall doe but juftice to name *Tycho, Hercules
the Second,* who releived his Predeceffor, long languifhing and
ready to faint under fo prodigious a burden ; which doubtlefs
was the Reafon, why *Kepler* called him, the *Modern Hipparchus.*

  And thus have we in a fhort Relation, rehearfed to you, what
we could gather together, concerning the Original, Progrefs, and
Advance of *Aftronomy,* from the higheft of times, of which there
remain any Authentick memorials, down to the deceafe of *Tycho
Brahe,* the Noble and the Great. As for what *Additions* this
excellent Science hath received, by the induftry of *Aftronomers*
in this prefent Age, by the help of the *Telefcope,* whofe Invention
may feem to have been unhappily deferred too long, as being deferred till fome Years after *Tychoes* death : they may be eafily
fummed up. For, all that our Dayes can juftly challenge the honour of difcovering, is (1.) the fpotts in the Sun: (2.) the inequality
of the fuperficies of the Moon: (3.) *Venus* fhifting her apparences,
as doth the Moon: (4.) *Mercury* and *Jupiter,* in fome Proportion,
doing the like : (5.) *Jupiter* with a kind of bound about him,
and guarded with four leffer Stars, as Attendants : (6.) *Saturn*
triple-bodied : (7.) the *Gallaxy* fully befet with fmall Stars :
and (8.) divers pale affemblies of very fmall Stars, feeming to be
only little white clouds in the *Welkin;* with fome other particulars lately remarked. Now, if you pleafe to add this to the former fummary : you have the whole (though brief) Story of *Aftronomy,* from its very infancy to that augmented ftate it now hath
attained to : I wifh I might have faid, to its *Full growth
and Perfection.* But, alas ! that is referved for
Pofterity.

www.ingramcontent.com/pod-product-compliance
Lightning Source LLC
Chambersburg PA
CBHW071400170526
45165CB00001B/125